Calculus Express

A Rapid Review and Formula Guide for the AP Calculus AB Exam

2nd Edition

Ryan Mettling

Performance Programs Company

6810 190th Street East

Bradenton, FL 34211

© 2026 by Performance Programs Company

Published by Performance Programs Company

Bradenton, Florida 34211

Printed in the United States of America

ISBN: 978-1965482421

Preface

Purpose and Benefits of Calculus Express

Calculus Express is a concise, student-friendly study guide designed to help you quickly and effectively prepare for the AP Calculus AB exam. Its streamlined approach also makes it an excellent companion for students in any introductory calculus course, where the content closely overlaps with AP Calculus AB.

To ensure maximum relevance, every topic is modeled directly from the College Board's official AP Calculus AB outline. The result is a focused, 110-page guide that covers all essential concepts, formulas, and examples—without the filler.

What sets Calculus Express apart is its brevity with precision. At just over 100 pages, it is specifically designed for efficient review and short-term mastery. You can actually cram with this book—memorize the key formulas and concepts, reinforce them with targeted examples, and walk into the exam confident and prepared. Other resources, often 600–900 pages long, function more like full-semester textbooks than exam-prep tools. They are valuable, but not optimized for rapid review.

If you want a fast, clean, no-nonsense study guide, Calculus Express delivers exactly that. No extraneous detail. No long derivations. Just the core formulas, definitions, examples, and strategies you need to succeed. As the author, I wrote the book I wish I had had when preparing for the AB exam.

Contents

Calculus Express is organized into five major sections:

- **Limits**
- **Derivatives**
- **Applications of Derivatives**
- **Integrals**
- **Applications of Integrals**

You'll begin with *limits*, the foundation of calculus—how to evaluate them, identify asymptotes, and apply the Intermediate Value Theorem.

From there, the book moves into *derivatives* and their applications. You'll learn derivative definitions, rules, rates of change, critical points, curve behavior, related rates, and more.

Next come *integrals* and their applications, including the Fundamental Theorem of Calculus, integration techniques, areas, accumulated change, and other AP-relevant uses.

The book concludes with 10 test taking tips according to Ryan.

About the Author:

Ryan Mettling, partner and publisher of Performance Programs Company, is an accomplished online curriculum designer, author, and course developer. He is responsible for the company's strategic planning, general management, printing and production, e-pub and retail platforms, and multi-channel marketing. Ryan is a co-author of the Real Estate License Exam Prep (RELEP) Series and Real Estate Math Express. Mr. Mettling is a member of the Real Estate Educators Association (REEA), and graduated Valedictorian from the University of Central Florida's College of Business Administration.

If you like Calculus Express, recommend it to your friends!

Table of Contents

Part I: Limits

1. Limit Definition

A limit is the y value that a function approaches at a specific x value.

A limit of a function exists if and only if the limit from the right equals the limit of from the left. A limit can exist at a hole.

The notation of a limit is the following:

as x → a   the limit of f(x)=   the limit of f(x)

$$x \to a+ \qquad\qquad x \to a-$$

The "+" sign indicates the limit from the right and the "−" sign indicates the limit from the left.

Example 1:
Limit Definition

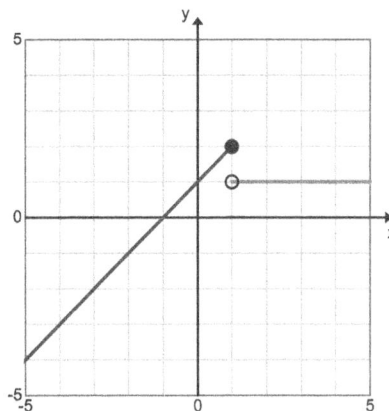

Limit x → 1+   =   1

Limit x → 1-   =   2

Limit x → 1   =   Ø (symbol for "does not exist")

2. Finding Limits

Some limits can be found by inserting the x value directly into the equation. However, in most problems more work is required. These challenging limits can be found through cancellation, rationalization, factorization, simplification of complex fractions, and by following L' Hôpital's rule. There will be an example of each method.

Example 2a:
Finding the Limit through Cancellation

Find the limit of $\dfrac{3x(x-1)}{x(x-2)}$ as $x \to 0$

Substituting zero in for x results in $\dfrac{0}{0}$ which is a hole. This is not what we are looking for. Thus, some other method is needed. The best method for this problem is to cancel the x's in front of the parentheses.

$$\text{Limit } x \to 0 \quad \frac{3x(x-1)}{x(x-2)}$$

Cancel:

$$\text{Limit } x \to 0 \quad \frac{3(x-1)}{x-2}$$

Now substitute in zero in for x to find the limit

$$\text{Limit } x \to 0 \quad \frac{3(0-1)}{0-2}$$

Simplify:

$$\frac{-3}{-2}$$

Simplify for final answer:

The limit as $x \to 0$ of $\dfrac{3x(x-1)}{x(x-2)}$ is $\dfrac{3}{2}$.

The graph of this function shows us the limit at x=0 is 3/2.

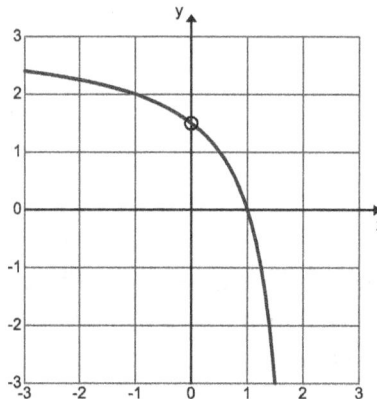

Example 2b:
Find the Limit through Rationalization

Find the limit of $\dfrac{(\sqrt{x+9})-2}{x+5}$ as x→ -5

With this limit question, there is nothing that can be cancelled. Therefore, another method must be utilized. Rationalization would be the most effective approach for this particular problem. The first step in rationalization is to find the conjugate of either the numerator or denominator and multiply the conjugate to both the numerator and denominator. A conjugate is a binomial formed by changing the sign of the second term (from positive to negative or negative to positive). Once completed, you will be able to cancel some portion of the equation. Then you can find the limit. So for this problem, we need the conjugate of the numerator. The conjugate is $(\sqrt{x+9})+2$.

$$\text{Limit as x→ -5} \quad \dfrac{(\sqrt{x+9})-2}{x+5}$$

<u>Multiply</u> both the numerator and denominator by the conjugate:

$$\text{Limit as x→ -5} \quad \dfrac{(\sqrt{x+9})-2}{x+5} \times \dfrac{(\sqrt{x+9})+2}{(\sqrt{x+9})+2}$$

$$\text{Limit x→ -5} \quad \dfrac{x+9+2(\sqrt{x+9})-2(\sqrt{x+9})-4}{(x+5)(\sqrt{x+9}+2)}$$

<u>Simplify</u>:

$$\text{Limit x→ -5} \quad \dfrac{(x+5)}{(x+5)(\sqrt{x+9}+2)}$$

Now <u>cancel out</u> the (x+5)'s and the limit can be found.

$$\text{Limit x→ -5} \quad \dfrac{1}{(\sqrt{x+9}+2)}$$

<u>Replace</u> x with -5:

$$\text{Limit x→ -5} \quad \dfrac{1}{(\sqrt{-5+9}+2)}$$

9

<u>Simplify</u>:

$$\text{Limit } x \rightarrow -5 \quad \frac{1}{(\sqrt{4}+2)}$$

$$\text{Limit } x \rightarrow -5 \quad \frac{1}{4}$$

The limit of $\dfrac{(\sqrt{x+9})-2}{x+5}$ as $x \rightarrow$ -5 is $\dfrac{1}{4}$.

Example 2c:
Find the Limit through Factorization

Find the limit of $\dfrac{x^2-4x+3}{x-3}$ as $x \rightarrow 3$

The most effective method for finding this problem's limit is to factor the numerator and see if anything will cancel with the denominator (x-3). The factor of the numerator is x-3 and x-1.

<u>Factor</u> the numerator:

$$\text{Limit as } x \rightarrow 3 \quad \frac{(x-1)(x-3)}{(x-3)}$$

<u>Cancel</u> out the x-3's:

$$\text{Limit as } x \rightarrow 3 \ (x-1)$$

<u>Replace</u> x with three to find the limit:

$$\text{Limit as } x \rightarrow 3 \quad (3-1) = 2$$

The limit of $\dfrac{x^2-4x+3}{x-3}$ as $x \rightarrow 3$ is 2.

Example 2d:
Find the Limit by Manipulating Complex Fractions

Find the limit of $\dfrac{\dfrac{1}{3+x}-\dfrac{1}{3}}{x}$ as $x \rightarrow 0$

The first step to find the limit is to simplify the numerator of the equation. This is accomplished by finding a common denominator of the two fractions.

Find the common denominator of numerator:

$$\text{Limit as } x \rightarrow 0 \quad \dfrac{\dfrac{-3-x+3}{3(3+x)}}{x}$$

<u>Simplify</u>:

$$\text{Limit as } x \rightarrow 0 \quad \dfrac{\dfrac{-x}{9+3x}}{x}$$

Now the denominator's reciprocal ($\dfrac{1}{x}$) can be multiplied by the numerator to simplify the equation further:

$$\text{Limit as } x \rightarrow 0 \quad \left(\dfrac{-x}{9+3x}\right) \times \left(\dfrac{1}{x}\right)$$

<u>Cancel</u> if possible (in this case the two solo x's):

$$\text{Limit as } x \rightarrow 0 \quad \left(\dfrac{-1}{9+3x}\right) \times \left(\dfrac{1}{1}\right)$$

<u>Simplify</u>:

$$\text{Limit as } x \rightarrow 0 \quad \left(\dfrac{-1}{9+3x}\right)$$

<u>Substitute</u> x with 0:

$$\text{Limit as } x \rightarrow 0 \quad \left(\dfrac{-1}{9+3(0)}\right)$$

<u>Simplify</u>:

$$\text{Limit as } x \to 0 \quad -\frac{1}{9}$$

The limit of $\dfrac{\dfrac{1}{3+x}-\dfrac{1}{3}}{x}$ as $x \to 0$ is $-\dfrac{1}{9}$.

Example 2e:
Find The Limit By Following L' Hôpital's Rule

L' Hôpital's Rule states that if a hole limit exists (when you substitute the stated value for x, BOTH the numerator and denominator are zero), the limit can be found by finding the derivative of both the numerator and denominator.

Find the limit of $\dfrac{x^2-9}{x-3}$ as $x \to 3$

This equation's limit can be found using L' Hôpital's Rule. So the next step is to find the derivative of the numerator and denominator.

$$\text{Limit as } x \to 3 \quad \frac{2x}{1}$$

<u>Simplify</u>:

$$\text{Limit as } x \to 3 \quad 2x$$

<u>Substitute</u> 3 for x:

$$\text{Limit as } x \to 3 \quad 6$$

The limit of $\dfrac{x^2-9}{x-3}$ as $x \to 3$ is 6.

3. *Vertical Asymptotes*

As the x values of a function approach a specific number, the y values increase exponentially to either positive or negative infinity on both sides of this specific x value. This is known as a vertical asymptote. For vertical asymptotes to exist, the denominator of any function must be zero and there must be any value other than zero in the numerator. The vertical asymptote is visible when graphing with a near vertical increase or decrease on both sides of the asymptote's x value. On many graphing devices, a line appears to go through the vertical asymptote to keep the graph connected as one "line." This line is not really there; it is just a way for graphing devices to represent the data without crashing.

Example 3:
Vertical Asymptotes

$F(x) = \dfrac{x^2 + 3x + 1}{16x^2 - 9}$. Find the x values where there are vertical asymptotes.

Set denominator to zero and <u>solve</u> for x:

$$16x\char`\^2-9=0$$

$$16x\char`\^2=9$$

$$x\char`\^2 = \frac{9}{16}$$

$$x = \sqrt{\frac{9}{16}}$$

$$x = \pm\,\frac{3}{4}$$

Check to see what the numerator equals. If it is anything but zero there will be a vertical asymptote.

$$x^2 + 3x + 1 \text{ at } x = \frac{3}{4}$$

$$\left(\frac{3}{4}\right)^2 + (3)\left(\frac{3}{4}\right) + 1$$

$$\left(\frac{3}{4}\right)^2 + (3)\left(\frac{3}{4}\right) + 1 = \frac{61}{16} \text{ (passes test)}$$

There is a vertical asymptote at x=.75.

$$x^2 + 3x + 1 \text{ at } x = -\frac{3}{4}$$

$$\left(-\frac{3}{4}\right)^2 + (3)\left(-\frac{3}{4}\right) + 1$$

$$(-\frac{3}{4})^2 + (3)(-\frac{3}{4}) + 1 = -\frac{11}{16} \text{ (passes test)}$$

There is a vertical asymptote at x=-.75

There will be a vertical asymptote at x= .75 and x= -.75 because the denominators at those x values will be zero and the numerators do not equal zero.

The graph below demonstrates this equation visually with the vertical asymptotes at $x=\pm\frac{3}{4}$.

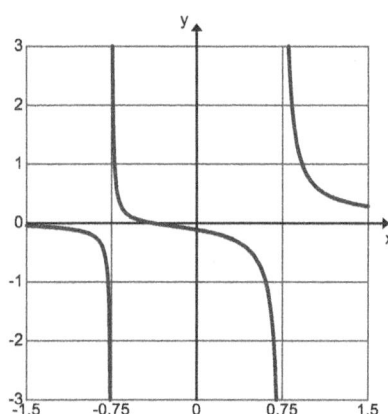

4. *End Behavior Model (Horizontal Asymptotes)*

As the x values of a function approach positive or negative infinity, the y values tend to get closer and closer to a specific number. This is known as a horizontal asymptote or as the end behavior model. To find a horizontal asymptote, select the highest x power of both the numerator and denominator and simplify just those two terms. The simplified answer is the value the equation will be approaching as $x \to \pm\infty$.

Example 4:
Horizontal Asymptotes

F(x)= $\dfrac{3x+2}{x-1}$ Find the horizontal asymptote as $x \to \pm\infty$

Select the highest x powers and remove other terms from the equation:

$$\frac{3x}{x}$$

14

Simplify:

$$\frac{3x}{x} = 3$$

As $x \to \pm\infty$ of $F(x) = \dfrac{3x+2}{x-1}$ the horizontal asymptote is 3.

The graph below shows this equation visually with the line of y=3 representing the horizontal asymptote. Also, x=1 is a vertical asymptote.

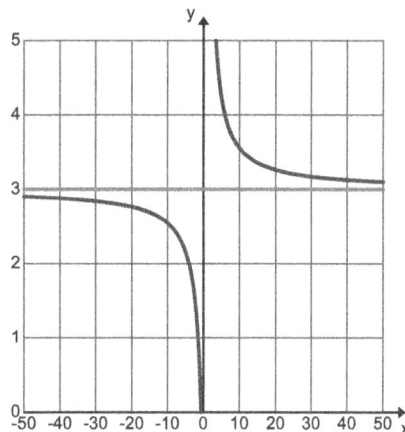

5. Continuity at a Point Rules

There are two rules regarding continuity at a point.

  a. For an interior point: y=f(x) is continuous at an interior point "c" of its domain if the limit of f(x)= f(c).

  b. For an endpoint: y=f(x) is continuous at a left endpoint "a" of its domain if the limit of f(x) as $x \to a^+$ = f(a). f(x) is also continuous at a right endpoint "b" of its domain if the limit of f(x) as $x \to b^-$ = f(b)

6. Making a Piecewise Function Continuous

The two most common types of piecewise function questions require you to find the value that will fill in a hole or a segment of the domain. This will connect the other segments and make the function continuous.

Example 6a:
Filling a Hole

Given: $f(x) = \begin{cases} \dfrac{\sin(x)}{x} & x \neq 0 \\ \underline{\quad ? \quad} & x = 0 \end{cases}$

Find the limit of sin(x)/x as x $\rightarrow$ 0

We need to find the value that will fill the hole at x=0. To do this, the limit of sin(x)/x as x $\rightarrow$ 0 needs to be found. L' Hôpital's rule can be used to find the limit of sin(x)/x as x $\rightarrow$ 0 (if you insert zero now, you get 0/0).

Compute the derivative of both the numerator and denominator of sin(x)/x as x $\rightarrow$ 0

$$\text{Limit as } x \rightarrow 0 \quad \frac{\cos(x)}{1}$$

Replace x with zero:

$$\text{Limit as } x \rightarrow 0 \quad \frac{\cos(0)}{1}$$

Solve:

$$\frac{\cos(0)}{1} = \frac{1}{1} = 1$$

The limit of sin(x)/x as x $\rightarrow$ 0 is 1.

Substitute one into the blank of the original problem so the hole is filled.

$$f(x) = \begin{cases} \dfrac{\sin(x)}{x} & x \neq 0 \\ 1 & x = 0 \end{cases}$$

Example 6b:
Filling a Gap in the Domain

Given: $f(x) = \begin{cases} 4x & x \le -1 \\ cx + d & -1 < x < 2 \\ -5x & x \ge 2 \end{cases}$

Find the values of c and d that assure that f(x) is continuous.

To fill the gap in the domain between -1<x<2, the values of c and d need to be found. Fortunately, this can be completed through relatively simple algebra.

Set the equation cx+d equal to one of the given equations at the x value they must meet for continuity. Start with 4x and cx+d. They must meet at x=-1.

cx+d=4x when x=-1

Substitute -1 for x:

-c+d=-4

Solve for either of the variables:

d=c-4

Now set the equation (cx+d) equal to the other equation (-5x) at the x value of the domain they must meet (x=2).

cx+d=-5x when x=2

Substitute c-4 for d:

cx+c-4=-5x when x=2

Substitute 2 for x and solve for c:

2c+c-4=-10

3c=-6

c=-2

Replace c with -2 in the d equation of d=c-4:

d=-2-4

d=-6

17

When c=-2 and d=-6, $f(x) = \begin{cases} 4x & x \le -1 \\ cx+d & -1 < x < 2 \\ -5x & x \ge 2 \end{cases}$ will be continuous.

7. Intermediate Value Theorem

The Intermediate Value Theorem (IVT) states if f(x) is a continuous function between [a,b] and f(a)=c and f(b)=d, then if i $\in [f(a), f(b)]$, then there is a k$\in$[a,b] such that f(k)=i.

This is a difficult concept to explain with words alone, so there is a graph displaying the Intermediate Value Theorem below.

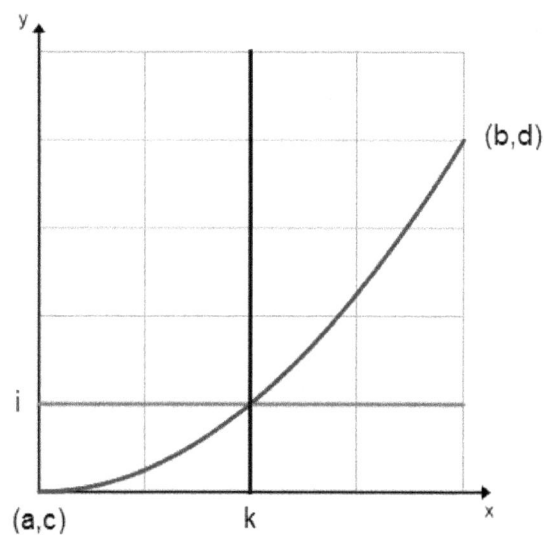

Part II: Derivatives

8. The Two Limit Definitions of a Derivative

a. The "A" Definition of a Derivative

A derivative of f(x) at x=a is the limit as $x \to a$ $\quad \dfrac{f(x)-f(a)}{x-a}$

Example 8a:
Using The "A Definitiion"

Find the derivative of $f(x) = \sqrt{x}$ as $x \to a$ using the "a" definition.

<u>Replace</u> f(x) and f(a) of $\dfrac{f(x)-f(a)}{x-a}$ with $\sqrt{x}$ and $\sqrt{a}$ respectively

Limit as $x \to a$ $\quad \dfrac{\sqrt{x}-\sqrt{a}}{x-a}$

<u>Rationalize</u> the equation

Limit as $x \to a$ $\quad \dfrac{\sqrt{x}-\sqrt{a}}{x-a}$ X $\dfrac{\sqrt{x}+\sqrt{a}}{\sqrt{x}+\sqrt{a}}$

<u>Multiply</u>

Limit as $x \to a$ $\quad \dfrac{x-(\sqrt{x}+\sqrt{a})+(\sqrt{x}+\sqrt{a})-a}{(x-a)(\sqrt{x}+\sqrt{a})}$

<u>Simplify</u>

Limit as $x \to a$ $\quad \dfrac{x-a}{(x-a)(\sqrt{x}+\sqrt{a})}$

<u>Cancel</u> the (x-a)'s

Limit as $x \to a$ $\quad \dfrac{1}{(\sqrt{x}+\sqrt{a})}$

<u>Substitute</u> "a" for x

Limit as $x \to a$ $\dfrac{1}{(\sqrt{a} + \sqrt{a})}$

Simplify

Limit as $x \to a$ $\dfrac{1}{2\sqrt{a}}$

The derivative of $f(x) = \sqrt{x}$ as $x \to a$ using the "a definition" is $\dfrac{1}{2\sqrt{a}}$

b. The "H" Definition of a Derivative

$f'(x) = \dfrac{f(x+h) - f(x)}{h}$ as the limit $h \to 0$

Example 8b:
Using the "H" Definition

Find the derivative of $f(x) = x^3$ using the "h" definition

Limit $h \to 0$ $\quad f'(x) = \dfrac{f(x+h) - f(x)}{h}$

Replace f(x) with x^3. Be careful with your substitution!

Limit $h \to 0$ $\quad f'(x) = \dfrac{(x+h)^3 - x^3}{h}$

Expand (x+h)^3

Limit $h \to 0$ $\quad f'(x) = \dfrac{(x^3 + 3x^2 h + 3xh^2 + h^3) - x^3}{h}$

Simplify by cancelling the x^3's

Limit $h \to 0$ $\quad f'(x) = \dfrac{3x^2 h + 3xh^2 + h^3}{h}$

Factor out h of the numerator

Limit $h \to 0$ $\quad f'(x) = \dfrac{h(3x^2 + 3xh + h^2)}{h}$

20

Cancel out the h in the numerator and denominator

$$\text{Limit } h \to 0 \quad f'(x) = 3x^2 + 3xh + h^2$$

Insert zero in for h

$$\text{Limit } h \to 0 \quad f'(x) = 3x^2 + 3x(0) + (0)^2$$

Simplify

$$\text{Limit } h \to 0 \quad f'(x) = 3x^2$$

The derivative of f(x)=x^3 using the "h" definition is $3x^2$.

9. *Instantaneous and Average Rate of Change*

The primary difference between the instantaneous rate of change (IROC) and the average rate of change (AROC) is that IROC relates to a specific point in time and AROC covers a period of time.

Example 9a:
Average Rate of Change

Given: A leaky pool is losing its water at a rate of $.3x^2 + x + 1$ gallons per hour.

Question A: What is the average rate of change between hours 7 through 9?

With AROC questions, look for the slope between the two stated endpoints. Thus for this question, the slope between f(9) and f(7) is the AROC.

First list the slope equation.

$$\text{AROC} = \frac{f(a) - f(b)}{a - b}$$

Replace a with 9 and b with 7 (the time frame of the problem)

$$\text{AROC} = \frac{f(9) - f(7)}{9 - 7}$$

Find f(9)

$$f(9) = .3x^2 + x + 1$$

$$f(9) = .3(9)^2 + 9 + 1$$

21

<u>Simplify</u>

$$f(9)= 24.3+9+1$$

$$f(9)=34.3 \text{ (this is total amount of water lost in the first 9 hours)}$$

<u>Find</u> f(7)

$$f(7)= .3x^2 + x + 1$$

$$f(7)= .3(7)^2 + 7 + 1$$

<u>Simplify</u>

$$f(7)= 14.7+7+1$$

$$f(7)= 22.7 \text{ (this is total amount of water lost in the first 7 hours)}$$

<u>Insert</u> 34.3 into AROC equation for f(9) and 22.7 for f(7)

$$\text{AROC}= \frac{34.3 - 22.7}{9 - 7}$$

<u>Simplify</u>

$$\text{AROC}= \frac{11.6}{2}$$

$$\text{AROC}= 5.8$$

The average rate of change of water lost between the seventh and ninth hour is 5.8 gallons per hour.

Example 9b:
Instantaneous Rate of Change

Given: A leaky pool began to lose its water at a rate of $.3x^2 + x + 1$ gallons per hour.

Question B: What is the instantaneous rate of change three hours in?

To find the IROC, find the derivative of the given equation and then insert the desired time period for x (3 for this question).

$$f(x)= .3x^2 + x + 1$$

Derive

$$f'(x) = .6x + 1$$

Substitute 3 into x

$$f'(3) = .6(3) + 1$$

$$f'(3) = 1.8 + 1$$

$$f'(3) = 2.8$$

The IROC of water lost at the third hour is 2.8 gallons per hour.

10. Calculating and Graphing Derivatives on TI-83 Calculators

The TI-83 series of graphing calculators (manufactured by Texas Instruments) is commonly owned by calculus students. On problems where calculators are allowed, knowing how to properly work a graphing calculator can save you time and allow you to double check your answers. However, it is crucial to know how to calculate derivatives by hand since many questions do not permit you to use a TI-83 or other similar device.

a. How to Find Specific Derivatives with the TI-83

Open "Y="

Enter the problem's function in " $y_1 =$ "

Graph the function

Enter "2nd mode" and push the "Trace" button (it has "calc" above it in yellow)

Select the 6th option which says "Dy/dx"

Enter the x value for which you want the derivative

Push "Enter" and answer appears in lower left hand corner.

b. How to Graph Derivatives with the TI-83

Open "Y="

Push the "Math" button

Select "8: nDeriv"

<u>Enter</u> the function and add ",x,x" after the function

<u>Push</u> the "Graph" button

11. The Six Trigonometry Derivatives and the Six Inverse Trigonometry Derivatives

a. The Trigonometric Derivatives

Y	Y'
sin(x)	cos(x)
cos(x)	-sin(x)
tan(x)	$\sec^2(x)$
cot(x)	$-\csc^2 x$
sec(x)	sec(x)tan(x)
csc(x)	-csc(x)cot(x)

b. The Inverse Trigonometric Derivatives

Y	Y'
arcsin(x)	$\dfrac{1}{\sqrt{1-x^2}}$
arccos(x)	$-\dfrac{1}{\sqrt{1-x^2}}$
arctan(x)	$\dfrac{1}{1+x^2}$
arccsc(x)	$-\dfrac{1}{|x|\sqrt{x^2-1}}$
arcsec(x)	$\dfrac{1}{|x|\sqrt{x^2-1}}$
arccot(x)	$-\dfrac{1}{1+x^2}$

24

12. Derivative Rules

a. The Power Rule

Let's say a function is $f(x) = x^n$

The derivative of $f(x) = x^n$ will be $f'(x) = nx^{n-1}$

This is known as the power rule and is the most commonly utilized method to find derivatives.

Example 12a:
The Power Rule

f(x)	$f'(x)$
5x	5
x	1
23	0
x^4	$4x^3$
$x^{\frac{1}{2}}$	$\frac{1}{2}x^{-\frac{1}{2}}$
$\frac{1}{x}$	$-x^{-2}$
$\frac{1}{x^2}$	$-2x^{-3}$
$3x^5 + 4x + 2$	$15x^4 + 4$
$45x^{-7}$	$-315x^{-8}$

b. The Product Rule

The product rule is necessary when the derivative of two or more functions multiplied together is needed.

Let's say a function is f(x)= (u)(v)

By using the product rule, the derivative of f(x)= (u)(v) is $f'(x) = u'v + v'u$

I would advise memorizing the product rule with the letters u and v. Those terms will be referenced in the following examples for clarity.

Example 12b-1:
The Product Rule

Find $f'(x)$ of f(x)= $(7x^2 + 3x)(x + 9)$

To find the derivative, the derivative of the "u" portion (7x^2+3x) needs to be found and then multiplied by the regular "v" portion (x+9).

Find the derivative of u (7x^2+3x)

> 14x+3

Multiply with v (x+9)

> (14x+3)(x+9)

This is $u'v$

Next, the derivative of the "v" portion (x+9) needs to be found and then multiplied by the original "u" portion (7x^2+3x).

Find the derivative of v (x+9).

> 1

Multiply 1 times u (7x^2+3x)

> (7x^2+3x)

This is $v'u$

Add together $u'v$ ((14x+3)(x+9)) and $v'u$ ((7x^2+3x))

$$\text{If f(x)} = (7x^2 + 3x)(x + 9),$$
$$f'(x) \text{ is } (14x + 3)(x + 9) + (7x^2 + 3x)$$

26

Example 12b-2:
The Product Rule

Find $f'(x)$ of f(x)= $(\sin(x)+3x^3)(\tan(x))$

To find the derivative, the derivative of the "u" portion (sin(x)+3x^3) needs to be found and then multiplied by the original "v" portion (tan(x)).

Find the derivative of u (sin(x)+3x^3)

$$\cos(x)+9x^2$$

Multiply with v (tan(x))

$$(\cos(x)+9x^2)(\tan(x))$$

This is $u'v$

Next, the derivative of the "v" portion (tan(x)) needs to be found and then multiplied by the original "u" portion (sin(x)+3x^3).

Find the derivative of v (tan(x))

$$\sec^2(x)$$

Multiply $\sec^2(x)$ times u (sin(x)+3x^3).

$$(\sec^2(x))(\sin(x)+3x^3)$$

This is $v'u$

Add together $u'v$ $((\cos(x)+9x^2)(\tan(x)))$ and $v'u$ $((\sec^2(x))(\sin(x)+3x^3))$

If f(x)= $(\sin(x)+3x^3)(\tan(x))$ then

$$f'(x) = (\cos(x)+9x^2)(\tan x)+(\sec^2(x))(\sin(x)+3x^3)$$

c. The Quotient Rule

The quotient rule is helpful when the derivative of a function divided by another function is needed.

Let's say a function is f(x)= $\dfrac{u}{v}$

By using the quotient rule, $f'(x) = \dfrac{u'v - v'u}{v^2}$

As with the product rule, I would advise memorizing the quotient rule with the letters u and v. I will refer to those terms in the following example.

Example 12c:
The Quotient Rule

Find $f'(x)$ of f(x)= $\dfrac{7x^4}{\cos(x)}$

The first step is to find $u'v$ where 7x^4 is u and cos(x) is v.

<u>Find</u> the derivative of u (7x^4)

$$28x^3$$

<u>Multiply</u> with v (cos(x))

$$(28x^3)(\cos(x))$$

This is $u'v$

Now find $v'u$

<u>Find</u> the derivative of v (cos(x))

$$-\sin(x)$$

<u>Multiply</u> with u (7x^4)

$$-(\sin(x))\,(7x\text{^}4)$$

This is $v'u$

<u>Find</u> v^2 by squaring v (cos(x)).

$$\cos^2(x)$$

28

This is v^2

Substitute the respective values for $u'v$, $v'u$, and v^2 into the quotient rule equation of $f'(x) = \dfrac{u'v - v'u}{v^2}$

If f(x)= $\dfrac{7x^4}{\cos(x)}$, $f'(x) = \dfrac{(28x^3)(\cos(x)) + (\sin(x))(7x^4)}{\cos^2(x)}$

d. The Chain Rule

The chain rule is used to find the derivative when the equation has "layers" of functions. The chain rule can also be referred to as differentiation of compositions of functions.

The process of the chain rule is: find the derivative of the outside function and multiply this derivative with the orginal inside function. This product is then multiplied by the derivative of the inside function.

This can also be shown with "f" as the outside function and "g" as the inside function.

f(g(x))

The derivative of f(g(x)) would be $(f'g(x))(g'(x))$

Example 12d-1:
The Chain Rule

Find the derivative of f(x)= $(2x + 3)^3$

In this example, the "^3" is the "outside" equation and 2x+3 is the "inside" function.

The derivative of anything raised to the third power is 3(_____)^2. Fill in the blank with 2x+3

3(2x+3)^2

This is the first half of what is needed, now the derivative of the inside function (2x+3) must be found.

f' of f(2x+3) is 2

Multiply 2 times 3(2x+3)^2 for the final answer.

The derivative of f(x)= $(2x + 3)^3$ is $6(2x + 3)^2$

29

Example 12d-2:
The Chain Rule

Find the derivative of f(x)= $\sin(\cos x)$

In this example, sin(__) is the outside equation and cos(x) is the inside equation.

The first step is to <u>find</u> the derivative of sin.

The derivative of sin is cos

<u>Multiply</u> cos with the interior equation which is cos(x)

cos(cos(x))

<u>Find</u> the derivative of the inside function (cos(x))

-sin(x)

<u>Multiply</u> –sin(x) by cos(cos(x)) for the final answer

The derivative of f(x)= $\sin(\cos x)$ is (cos(cos(x))(-sin(x)).

e. Implicit Differentiation

Implicit differentiation is necessary when the function y is written implicitly as a function of x. For example 8y= sin(x) and $y^5 = 7x - 4$ would need to be solved through implicit differentiation. There are four steps to solve implicit differentiation problems.

Step 1 - Differentiate both sides of the equation with respect to x.

Step 2 - Collect the terms with $\dfrac{dy}{dx}$ on one side of the equation.

Step 3 - Factor out $\dfrac{dy}{dx}$

Step 4 - Solve for $\dfrac{dy}{dx}$

Example 12e-1:
Implicit Differentiation

Find the derivative of $y^2 = x$

Step 1 - <u>Differentiate</u> both sides of the equation with respect to x.

$$2y\frac{dy}{dx} = 1$$

Step 2 - <u>Collect</u> the terms with $\frac{dy}{dx}$ on one side of the equation.

$$\frac{dy}{dx} = \frac{1}{2y}$$

Step 3 was not needed for this particular problem.

The derivative of $y^2 = x$ is $\frac{dy}{dx} = \frac{1}{2y}$

Example 12e-2:
Implicit Differentiation

Find the derivative of $2y = x^2 + \sin(y)$

Step 1 - <u>Differentiate</u> both sides of the equation with respect to x.

$$2\frac{dy}{dx} = 2x + (\cos(y))\frac{dy}{dx}$$

Step 2 - <u>Collect</u> the terms with $\frac{dy}{dx}$ on one side of the equation.

$$2\frac{dy}{dx} - (\cos(y))\frac{dy}{dx} = 2x$$

Step 3 - <u>Factor out</u> $\frac{dy}{dx}$

$$(2\text{-cos}(y))(\frac{dy}{dx}) = 2x$$

Step 4 - <u>Solve</u> for $\frac{dy}{dx}$

$$\frac{dy}{dx} = \frac{2x}{2 - \cos(y)}$$

The derivative of $2y = x^2 + \sin(y)$ is $\frac{dy}{dx} = \frac{2x}{2 - \cos(y)}$

13. Finding Tangent and Normal Lines

To show how to find the tangent and normal line at a point, the example of $y = x^3$ and the point (1,1) will be utilized.

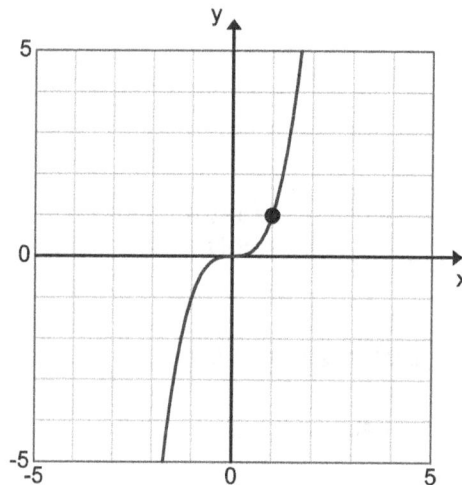

Part A- Finding the Tangent Line

The tangent line touches the function once at the point we focusing on (1,1) and has the same slope as the IROC at that point. The first step in finding the tangent line is to find the slope at the desired x value.

<u>Find</u> the derivative of $y = x^3$

$$y' = 3x^2$$

<u>Find</u> the slope when x=1

$$f'(1) = 3(1)^2$$

$$f'(1) = 3$$

Use the point slope equation $y - y_1 = m(x - x_1)$ with 3 as the slope (m), and (1,1) as the point to <u>find</u> the tangent line.

y-1=3(x-1)

y-1=3x-3

y=3x-2

The tangent line of $y = x^3$ at the point (1,1) is y=3x-2

This tangent line has been added to the graph of the original function ($y = x^3$) below.

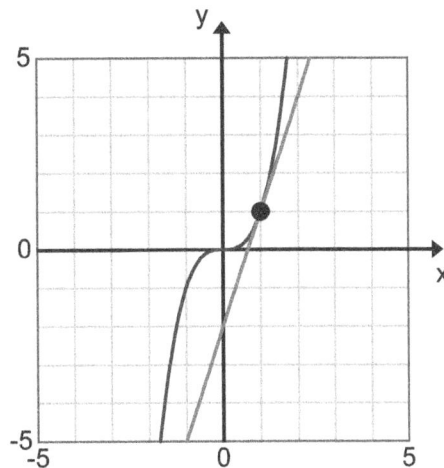

Part B- Finding the Normal Line

The only difference between the tangent line and the normal line is that to find the normal line, you need the negative reciprocal of the slope at the desired point instead of the regular slope. Remember from the tangent line example, that the slope at (1,1) was 3. So for the normal line the slope will be $-\dfrac{1}{3}$.

Use the point slope equation $y - y_1 = m(x - x_1)$ with $-\dfrac{1}{3}$ as the slope and the point as (1,1).

33

$$y-1=-\frac{1}{3}(x-1)$$

$$y-1=-\frac{1}{3}x+\frac{1}{3}$$

$$y=-\frac{1}{3}x+\frac{4}{3}$$

The normal line of $y=x^3$ at the point (1,1) is $y=-\frac{1}{3}x+\frac{4}{3}$

This normal line has been added to the graph of the original function ($y=x^3$) below.

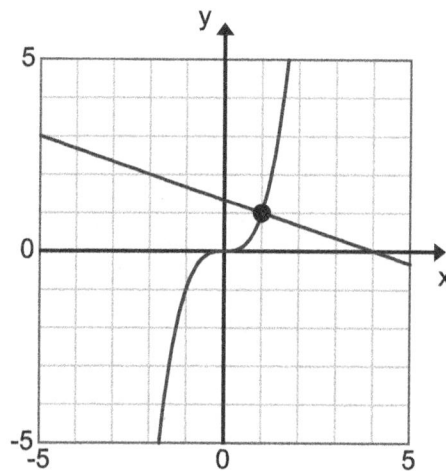

14. Absolute Value Function

With absolute value functions, the entire domain will result in positive y values. The two horizontal bars around the function indicate that it is an absolute value function.

Example 14:
Absolute Value Function

$$y=|2x-2|$$

34

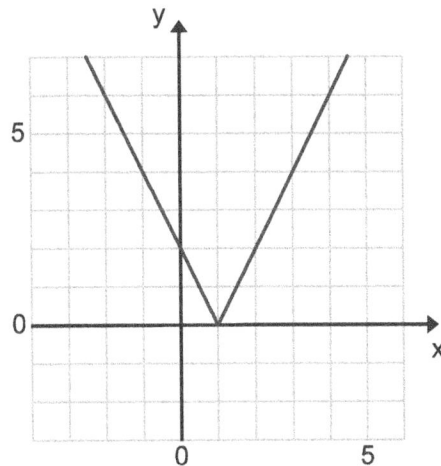

All the y values are equal or greater than zero. Normally, a non-exponential function would be a straight line but absolute value functions make a sharp turn upward like an "elbow" at y=0. The slope of $y = |2x - 2|$ is 2 when $x \geq 1$ and the slope is -2 when $x < 1$.

Another way to write an absolute value functions is as a piecewise function.

$y = |2x - 2|$ as a piecewise function $\begin{cases} 2x - 2, x \geq 1 \\ -2x + 2, x < 1 \end{cases}$

15. Derivative of a Function Raised to Another Function

Example 15:
Derivative of a Function Raised to Another Function

Find the derivative of $y = x^x$

The first step is to find the natural logarithm(ln) of both sides of the equation.

ln(y)=ln(x^x)

You can rewrite ln(x^x) as xln(x). This is why you want the natural logarithm.

ln(y)=xln(x)

Differentiate both sides. The derivative of ln(x) is 1/x. The chain rule is utilized in this example for xln(x).

$$(\frac{1}{y})(\frac{dy}{dx}) = (1)(\ln(x)) + (\frac{1}{x})(x)$$

Simplify

$$(\frac{1}{y})(\frac{dy}{dx}) = \ln(x) + 1$$

Move the 1/y over to the right side of the equation so dy/dx is by itself.

$$\frac{dy}{dx} = y(\ln(x) + 1)$$

Since we know from the original equation that $y = x^x$, we can substitute x^x for y.

$$\frac{dy}{dx} = x^x(\ln(x) + 1)$$

Note: $\frac{dy}{dx}$ is the same as y'

The derivative of $y = x^x$ is $x^x(\ln(x) + 1)$

Part III: Applications of Derivatives

16. How to Find Where a Function is Increasing, Decreasing, Concave Up, Concave Down, and a Function's Points of Inflection.

The following equation will be used for the rest of this section.

$$f(x) = -x^4 + 4x^3 - 4x + 1$$

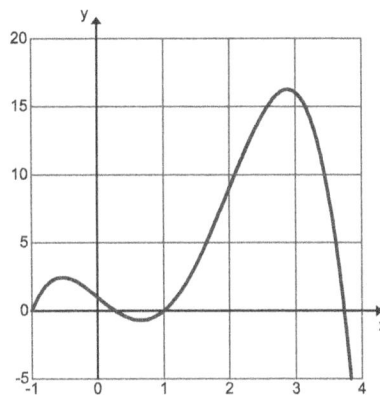

a. Finding Where the Function is Increasing or Decreasing.

The first step to find the behavior of the function, $f(x) = -x^4 + 4x^3 - 4x + 1$, is to calculate its derivative.

$$y' = -4x^3 + 12x^2 - 4$$

To find where the function is increasing or decreasing, set the derivative equal to zero and find the values of x where the derivative equals zero.

$$0 = -4x^3 + 12x^2 - 4$$

Solve for x (easiest with a graphing calculator)

$0 = -4x^3 + 12x^2 - 4$ when x= -.532, .653, or 2.879

With these x values, create a signline similar to what is at the top of the next page.

$$f' \leftarrow \text{-----------}|\text{-----------------}|\text{-----------------}|\text{-------------} \rightarrow$$

$$\text{-.532} \qquad \text{.653} \qquad \text{2.879}$$

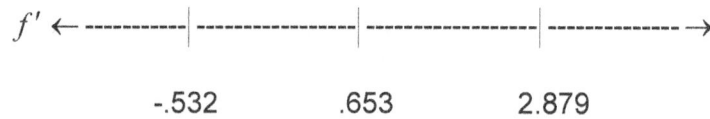

Now find whether f' is positive or negative between the four gaps in the signline by testing a potential value of f'. A negative slope means the function is decreasing; a positive slope means the function is increasing. Any number between the "gaps" will be sufficient, so I would advise selecting the less time-consuming ones (such as whole numbers or zero if possible).

$$y' = -4x^3 + 12x^2 - 4$$

For the first gap, $y'(-1) = -4(-1)^3 + 12(-1)^2 - 4$

$$y'(-1) = 12$$

12 is a positive number, thus the function is increasing in that gap of the signline.

For the second gap, $y'(0) = -4(0)^3 + 12(0)^2 - 4$

$$y'(0) = -4$$

-4 is a negative number, thus the the function is decreasing in this range of the domain.

For the third gap, $y'(1) = -4(1)^3 + 12(1)^2 - 4$

$$y'(1) = 4$$

Thus the function is increasing in that segment of the domain.

For the final gap, $y'(3) = -4(3)^3 + 12(3)^2 - 4$

$$y'(3) = -4$$

The function is decreasing in that segment of the signline.

Now you can display where the function is increasing or decreasing on the signline.

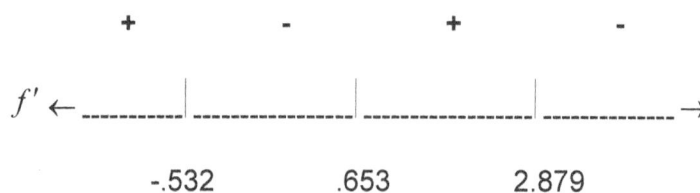

$$\qquad + \qquad\qquad - \qquad\qquad + \qquad\qquad -$$

$$f' \leftarrow \text{----------}|\text{----------------}|\text{----------------}|\text{-------------} \rightarrow$$

$$\text{-.532} \qquad \text{.653} \qquad \text{2.879}$$

The function $y = -x^4 + 4x^3 - 4x + 1$ is increasing from $(-\infty, -.532)$ and from (.653, 2.879)

The function $y = -x^4 + 4x^3 - 4x + 1$ is decreasing from (-.532, .653) and from (2.879, ∞)

b. Concavity and Points of Inflection

Concavity describes the behavior of the slope. When concavity is positive, the slope is increasing. When concavity is negative, the slope is decreasing. When concavity is zero, the slope is constant.

Concavity is found through the same process as the behavior of the function, except for you start by calculating the derivative of the derivative instead of the derivative of the original equation. Thus, to find concavity for this example, the first step is to find the derivative of $y' = -4x^3 + 12x^2 - 4$.

$$y'' = -12x^2 + 24x$$

Set the second derivative equal to zero and solve for x.

$$0 = -12x^2 + 24x \text{ when x= 0 or 2}$$

With these x values create a signline as in part a.

$$f'' \leftarrow \text{-----------|-----------------|-------------} \rightarrow$$

$$\qquad\qquad 0 \qquad\qquad 2$$

Now find whether f'' is positive or negative between the three segments in the signline by substituting a potential value into f''. A negative number means the function is concave down; a positive number means the function is concave up.

$$y'' = -12x^2 + 24x$$

$$y''(-1) = -12(-1)^2 + 24(-1)$$

$y''(-1) = -36$, thus this segment of the signline is concave down.

$$y''(1) = -12(1)^2 + 24(1)$$

$y''(1) = 12$, this range of the domain is concave up.

39

$$y''(3) = -12(3)^2 + 24(3)$$

$y''(3) = -36$, the last gap of the domain is concave down.

This information can now be placed on the signline.

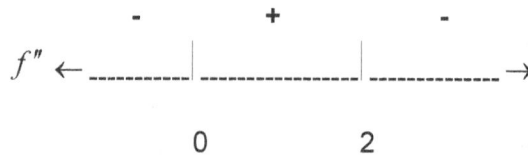

$$
\begin{array}{ccc}
- & + & - \\
f'' \leftarrow \text{_____}|\text{_____}|\text{_____} \rightarrow \\
0 \qquad\qquad 2
\end{array}
$$

The function $y = -x^4 + 4x^3 - 4x + 1$ is concave up from (0,2) and is concave down from $(-\infty,0)(2,\infty)$.

Points of inflection (POI's) occur at the x values where concavity either changes from concave up to concave down or from concave down to concave up. The POI's for $y = -x^4 + 4x^3 - 4x + 1$ are at x=0 and x=2. The signline is helpful to quickly see POI's.

17. The Second Derivative Test for Local Extrema

The second derivative test has two rules to help determine if a point is a relative maximum or relative minimum.

1. If f'(c)=0 and $f''(c) < 0$, then the function has a local maximum at x=c.

2. If f'(c)=0 and $f''(c) > 0$, then the function has a local minimum at x=c.

Example 17:
The Second Derivative Test

Find the extreme values of f(x)= $x^3 - 12x - 5$

<u>Find</u> the derivative of f(x)= $x^3 - 12x - 5$

$$f'(x) = 3x^2 - 12$$

<u>Find</u> the x values where f' is zero. Factoring out three will help here. This step will find the critical values that are either local maximums or minimums.

$$f'(x) = 3(x^2 - 4)$$

$$0 = 3(x^2 - 4)$$

$0 = 3(x^2 - 4)$ when x=2 or x=-2

Find the 2nd derivative of f(x)= $x^3 - 12x - 5$ by calculating the derivative of $f'(x) = 3x^2 - 12$

$$f''(x) = 6x$$

We need to test the critical values of x=2 and x=-2 to see if they are positve or negative for the equation $f''(x)$ = 6x. As stated by the 2nd derivative rule, if $f''(x)$ <0 the point is a local maximum. If $f''(x)$ >0, it is a local minimum.

$$f''(2) = 6(2)$$

$$f''(2) = 12$$

Thus f(x) has a local minimum at x=2

$$f''(-2) = 6(-2)$$

$$f''(-2) = -12$$

Thus f(x) has a local maximum at x=-2.

18. How To Find Local And Absolute Extrema Of A Function With A Limited Domain

As seen in previous problems, the way to find critical values is to set the derivative of a function equal to zero and solve for x. If the derivative changes from a positive number to a negative one, the extrema (critical value) is a local maximum. If the derivative changes from a negative number to a positive one, the extrema is a local minimum.

The absolute extrema is the x value that results in the highest y value (absolute maximum) or the x value that results in the lowest y value (absolute minimum) for the entire equation. To find absolute extrema, you have to test all the critical values AND endpoints (if the domain is limited) by substituting the critical values into the original equation and evaluating their y values.

For the following example, the domain of the function has been limited.

Example 18:
Find Local and Absolute Extreme

$$f(x) = 3x^3 - 9x^2 + 3x - 1.$$

The domain of the equation is $4 \geq x \geq -1$. Find the absolute maximum and minimum of the function.

The first step is to <u>find</u> the equation's derivative.

$$f'(x) = 9x^2 - 18x + 3$$

<u>Set</u> f' equal to zero and <u>find</u> the critical values. Note: the quadratic formula ($x = \dfrac{-b \pm \sqrt{b^2 - 4ac}}{2a}$) was utilized to find the values below.

$$0 = 9x^2 - 18x + 3$$

$$0 = 9x^2 - 18x + 3 \text{ when x=.184 or x=1.816}$$

Now create a signline of the derivative. x=-1 and x=4 will be endpoints of the signline since they are endpoints of the equation's domain.

$$f' \;\; |\text{---------------}|\text{-----------------}|\text{--------------}|$$

$$ \text{-1} \qquad \text{.184} \qquad \text{1.816} \qquad \text{4}$$

Now <u>find</u> whether f' is positive or negative between the three gaps in the signline by testing a potential value.

For the first gap,

$$f'(0) = 9(0)^2 - 18(0) + 3$$

$f'(0) = 3$; thus the function is increasing between $-1 \leq x \leq .184$

For the second gap,

$$f'(1) = 9(1)^2 - 18(1) + 3$$

$f'(1) = -6$; thus the function is decreasing between $.184 < x \leq 1.816$

For the third gap,

$$f'(2) = 9(2)^2 - 18(2) + 3$$

42

$f'(2) = 3$; the function is increasing between $1.816 < x \leq 4$

Update the signline.

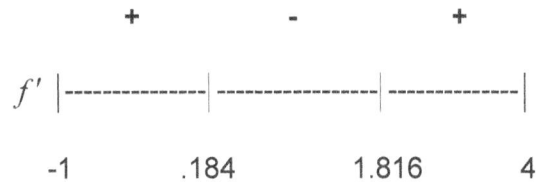

$$\begin{array}{ccccc} & + & - & + & \\ f' & |\text{---------------}| & \text{-----------------}| & \text{-------------}| \\ & \text{-1} & .184 & 1.816 & 4 \end{array}$$

Remember from earlier, if the derivative changes from a positive number to a negative one, the extrema (critical value) is a local maximum. If the derivative changes from a negative number to a positive one, the extrema is a local minimum.

Based on this information and the signline above, we know for sure that x=.184 is a local maximum and x=1.816 is a local minimum. However, we need to test these critical values along with the endpoints to determine the absolute extrema.

Test the endpoints (x=-1,4) and the critical values (x=.184, 1.816) by substituting these values back into original equation of $f(x) = 3x^3 - 9x^2 + 3x - 1$.

$$f(-1) = 3(-1)^3 - 9(-1)^2 + 3(1) - 1$$

$$f(-1) = -16$$

$$f(.184) = 3(.184)^3 - 9(.184)^2 + 3(.184) - 1$$

$$f(.184) = -.734$$

$$f(1.816) = 3(1.816)^3 - 9(1.816)^2 + 3(1.816) - 1$$

$$f(1.816) = -7.266$$

$$f(4) = 3(4)^3 - 9(4)^2 + 3(4) - 1$$

$$f(4) = 59$$

$f(4) = 59$ is the highest y value and is the absolute maximum of $f(x) = 3x^3 - 9x^2 + 3x - 1$ when the domain of the equation is $4 \geq x \geq -1$.

$f(-1) = -16$ is the lowest y value. That means f(-1) is the absolute minimum of $f(x) = 3x^3 - 9x^2 + 3x - 1$ when the domain of the equation is $4 \geq x \geq -1$..

This example demonstrated that without testing the endpoints, the correct answer would not be found.

19. Maximizing or Minimizing Problems

Example 19a:
Maximizing

1000 feet of chain fencing will be used to construct 6 cages in a design shown below. There are 8 total y length segments and 9 total x length segments necessary to make this design. Find the dimensions (values of x and y) which will maximize the enclosed area.

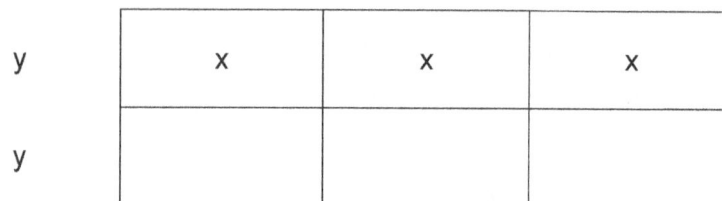

y	x	x	x
y			

The first step is to set up the equation of total fencing segments equal to the 1000 feet that can be used. There will be 9 "x" length and 8 "y" length segments.

$$1000=9x+8y$$

Now <u>solve</u> for one of the variables

$$8y=-9x+1000$$

$$y=-\frac{9}{8}x+125$$

Next, provide the area equation for the design since we are trying to maximize enclosed area.

$$A=(3x)(2y)$$

<u>Substitute</u> $-\frac{9}{8}x+125$ for y

$$A=(3x)(2(-\frac{9}{8}x+125))$$

$$A=(3x)(-\frac{9}{4}x+250)$$

44

Continue to simplify.

$$A = -\frac{27}{4}x^2 + 750x$$

To maximize the area, you need to differentiate the area. Then, discover the critical values by finding where A' is zero.

$$A' = -\frac{27}{2}x + 750$$

$$0 = -\frac{27}{2}x + 750$$

$$\frac{27}{2}x = 750$$

$$x = 55.556$$

Create a signline to make sure that is the critical value that maximizes area.

$$A' \leftarrow \text{----------}|\text{-------------} \rightarrow$$

$$55.556$$

Now find whether A' is positive or negative between the two gaps in the signline. Insert a potential value into A' for each segment.

$$A'(55) = -\frac{27}{2}(55) + 750$$

$$A'(55) = 7.5$$

Thus the total area is increasing when $x \leq 55.556$

$$A'(56) = -\frac{27}{2}(56) + 750$$

$$A'(56) = -6$$

Thus the total area is decreasing when $x > 55.556$

This proves that the enclosed area is maximized when x=55.556. Now that we have x, all what remains is to substitute 55.556 for x in the equation $y = -\frac{9}{8}x + 125$.

$$y = -\frac{9}{8}(55.556) + 125$$

$$y = 62.5$$

Add the word "feet" back to those values to properly answer the question.

The dimensions that will maximize the enclosed area are y=62.5 feet and x=55.556 feet. The outer dimensions of the 3x by 2y will be 166.66 ft (3x) by 125 ft (2y).

y=62.5 feet	x=55.556 ft	x=55.556 ft	X=55.556 ft
y=62.5 feet			

Example 19b:
Minimizing

Suppose that the cost to manufacture a product is represented by the function $c(x) = x^3 - 6x^2 + 15x$. x represents 1,000's of units. Find the production level that minimizes average cost.

Hint: look where average cost equals marginal cost

Marginal cost: $c'(x) = 3x^2 - 12x + 15$

Average cost: $\frac{c(x)}{x} = x^2 - 6x + 15$

As stated in the provided hint, set average cost equal to marginal cost.

$$3x^2 - 12x + 15 = x^2 - 6x + 15$$

Move all the variables to one side of the equation and set the other side equal to zero.

$$2x^2 - 6x = 0$$

46

Factor out 2x

$$2x(x-3)=0$$

The marginal cost equals the average cost at x=3 or x=0.

x=0 is not a possible production number so it can be ignored.

Since the question is asking to minimize average cost, we must test x=3 with the average cost function ($\frac{c(x)}{x}$) to find if 3,000 units is the production value that minimizes cost. The 2nd derivative test is the easiest way to check.

If $\frac{c'(x)}{x}$ when x(3)=0 and $\frac{c''(x)}{x}$ when x(3)>0, then the function has a local minimum at x=3.

$$\frac{c'(x)}{x}=2x-6$$

$$\frac{c'(3)}{x}=2(3)-6$$

$$\frac{c'(3)}{x}=0$$

The first part of the test has been confirmed.

$$\frac{c''(x)}{x}=2$$

$$\frac{c''(3)}{x}=2$$

The second part of the test has been confirmed.

$\frac{c'(x)}{x}$ when x=3 is equal to 0 and $\frac{c''(x)}{x}$ when x=3 is greater than zero (2), thus the function has a minimum at x=3.

The production level that minimizes average cost is 3,000 units.

47

20. Related Rates

The formula for the volume of a cone is $V(t) = \frac{1}{3}\pi(r^2)h$. R is the radius and h is the height. Find the rate of change of the volume if $\frac{dr}{dt}$ = 2 inches per minute, h=3r, and r=6 inches. $\frac{dr}{dt}$ means the radius is increasing and can also be interpreted as r'.

The first step is to <u>substitute</u> h with 3r in the volume equation.

$$V(t) = \frac{1}{3}\pi(r^2)(3r)$$

<u>Simplify</u>

$$V(t) = \pi(r^3)$$

Differentiate this volume equation. This problem requires implicit differentiation. Do not forget $\frac{dr}{dt}$!

$$\frac{dv}{dt} = \pi(3r^2)(\frac{dr}{dt})$$

Now <u>substitute</u> dr/dt with 2 and <u>replace</u> r with 6 as given in the original problem.

$$\frac{dv}{dt} = \pi(3(6)^2)(2)$$

$$\frac{dv}{dt} = 216\,\pi\,\frac{inches^2}{minute}$$

The rate of change of the volume, if $\frac{dr}{dt}$ =2 inches per minute, h=3r, and r=6 inches, is an increase of $216\,\pi\,\frac{inches^2}{minute}$.

Example 20b:
Related Rates

An airplane pilot spots two cars converging on a point. One car is 15 miles to the east of the convergence point while traveling due west at 45 miles per hour. The other car is 20 miles north of convergence the point and is moving south at 60 miles per hour. At what rate is the distance between the cars decreasing?

To better understand this problem, a diagram is below.

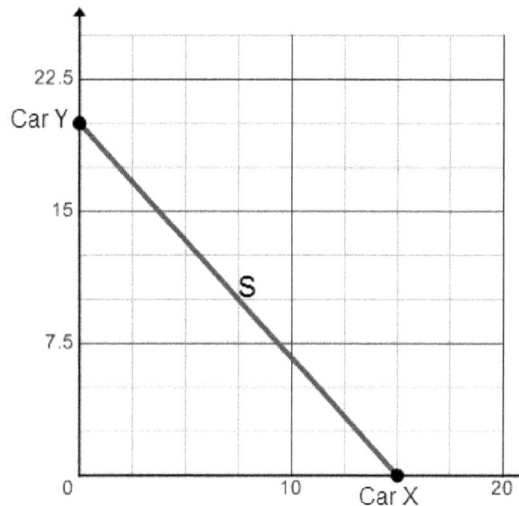

Y represents the car 20 miles from the point (0,0) with the $\dfrac{dy}{dt}$ of -60

X represents the car 15 miles from the point (0,0) with the $\dfrac{dx}{dt}$ of -45

The blue line S represents the distance between the two cars. $\dfrac{ds}{dt}$ is the rate which the distance between the cars is changing and is what the problem is ultimately trying to solve.

The first step is to find the distance s by applying the Pythagorean theorem. The Pythagorean theorem can be implemented since we have a right triangle.

$$s^2 = x^2 + y^2$$

We know what x and y are (15 and 20 respectively), so substitute those values in and solve for s.

$$s^2 = 15^2 + 20^2$$

49

$$s^2 = 225 + 400$$

$$s^2 = 625$$

$$s = \sqrt{625}$$

$$s = 25$$

Now that s has been found, <u>find</u> $\dfrac{ds}{dt}$ (the rate at which the distance between the cars decreasing) through implicit differentiation of the pythagorean theorem.

$$s^2 = x^2 + y^2$$

<u>Calculate</u> the derivative

$$2s\left(\dfrac{ds}{dt}\right) = 2x\left(\dfrac{dx}{dt}\right) + 2y\left(\dfrac{dy}{dt}\right)$$

Now <u>substitute</u> the known values (all known except for ds/dt)

$$2(25)\left(\dfrac{ds}{dt}\right) = 2(15)(-45) + 2(20)(-60)$$

<u>Solve</u> for ds/dt

$$(50)\left(\dfrac{ds}{dt}\right) = -1350 + (-2400)$$

$$(50)\left(\dfrac{ds}{dt}\right) = -3750$$

$$\dfrac{ds}{dt} = -75$$

The distance between the cars is decreasing at a rate of 75 miles per hour.

21. The Mean Value Theorem

According to the Mean Value Theorem (MVT), if f(x) is continous on [a,b], and differentiable on [a,b] then there must be a point c (a<c<b) where $\dfrac{f(b) - f(a)}{b - a} = f'(c)$

Example 21:
The Mean Value Theorem

Find the point of $y = x^2$ that satisfies the MVT when the domain is [-2,3].

The first step is to find the slope between f(-2) and f(3) of the function ($y = x^2$) using the equation $\dfrac{f(b) - f(a)}{b - a}$. For this problem, f(3) will be b and f(-2) will be a.

$$\frac{f(3) - f(-2)}{3 - (-2)}$$

f(3) is 9 and f(-2) is 4

Substitute these values into the numerator.

$$\frac{9 - 4}{3 - (-2)}$$

Simplify

$$\frac{5}{5} = 1$$

The average slope between [-2,3] is one. Now find the x values of the domain where the instantaneous rate of change (IROC) is one. To find these values, differentiate the original equation ($y = x^2$).

$$y = x^2$$

$$y' = 2x$$

Set y' equal to the slope found in the first part of the question (one in this case) and solve for x.

$$1 = 2x$$

$$x = \frac{1}{2}$$

Insert this value back into the orginal equation to find its corresponding y value.

$$y = (\frac{1}{2})^2$$

$$y = \frac{1}{4}$$

The point $(\frac{1}{2}, \frac{1}{4})$ satisfies the MVT of $y = x^2$ with a domain of [-2,3].

This example is demonstrated with the graph below. Note how two of the lines are parallel. The slope of the top parallel line was found in the first half of the problem and the point $(\frac{1}{2}, \frac{1}{4})$ (black dot on graph) was found in the second half.

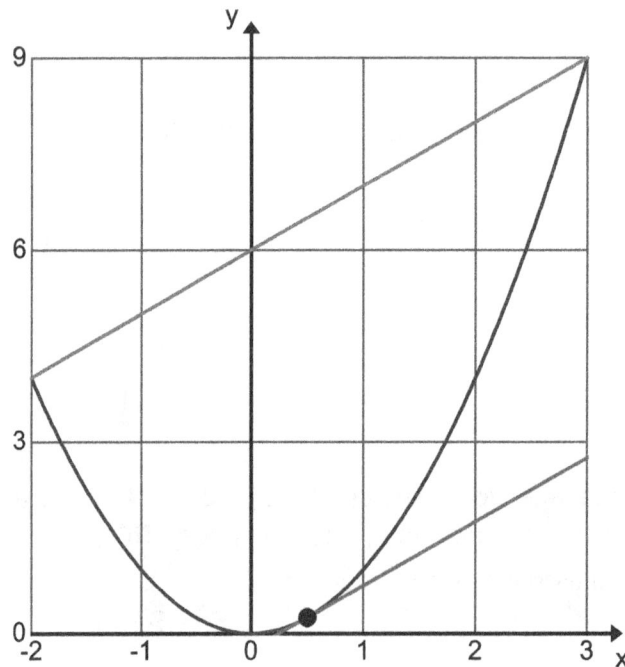

Part IV: Integration

22. Riemann Sums

Riemann sums are utilized to <u>estimate</u> the total area underneath a curve. This area under a curve is known as an integral. Riemann sums have a larger margin of error when compared to the other integration techniques covered in this unit. The most common Riemann sum varieties are left-hand, right-hand, midpoint.

a. Left-Hand Riemann Sums

Example 22a-1:
Left-Hand Riemann Sum

Find the area if $f(x) = \dfrac{1}{x}$,   a=1   b=3 LH_4

The first step to solve this problem is find the value of the intervals. The value of the intervals can be found by the equation $\dfrac{b-a}{n}$. N is the number in the bottom right of $LH_n, RH_n,$ or M_n. For this problem, n is 4. The values for a and b are given.

$$\frac{b-a}{n}$$

$$\frac{3-1}{4}$$

$\dfrac{1}{2}$ is the interval we will be using for this problem.

The formula for left-hand Riemann sums is set up below. Use the variable "i" to represent the intervals (i=.5 for this problem)

$$(i)[f(a)+f(a+i)+f(a+2i)+f(a+3i)+\ldots\ldots]$$

The total number of f(___) depends on n. For this problem there will be 4 different values of the function to find. Remember, a is 1 and i is .5.

$$(\frac{1}{2})[f(1)+f(1.5)+f(2)+f(2.5)]$$

Now find the values for f(_)

$$f(1) = \frac{1}{1} = 1$$

$$f(1.5) = \frac{1}{1.5} = \frac{2}{3}$$

$$f(2) = \frac{1}{2}$$

$$f(2.5) = \frac{1}{2.5} = \frac{2}{5}$$

Insert these values into the equation ((i)[f(a)+f(a+i)+f(a+2i)+f(a+3i)+......]).

$$(\frac{1}{2})[1 + \frac{2}{3} + \frac{1}{2} + \frac{2}{5}]$$

Simplify:

$$(\frac{1}{2})[2.567]$$

1.2835

The accumulated area found of $f(x) = \frac{1}{x}$, a=1 b=3 LH_4 is 1.2835. A graph of this Riemann Sum is below. The function f(x)= 1/x is the curved line and the total area found (1.2835) through four left-hand Riemann sums is the shaded area. Notice how the left-hand Riemann sums found more area than the actual area under the function.

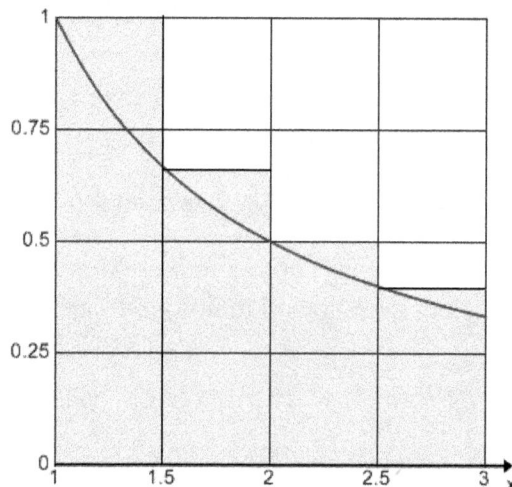

54

Example 22a-2:
Left-Hand Riemann Sum

Find the area if $f(x) = x^2 - x + 3$ a=0   b=3 LH_6

The first step to solve this problem is find the value of the intervals. They can be found by the equation $\dfrac{b-a}{n}$. N is 6.

$$\frac{b-a}{n}$$

$$\frac{3-0}{6}$$

$\dfrac{1}{2}$ is the interval for this problem.

<u>Insert</u> the appropiate information into the LH equation of
(i)[f(a)+f(a+i)+f(a+2i)+f(a+3i)+......] where i=1/2. There are 6 total f(__)'s.

$$(\frac{1}{2})[f(0) + f(.5) + f(1) + f(1.5) + f(2) + f(2.5)]$$

Now find the correct values for f(__)

$$f(0) = (0)^2 - (0) + 3 = 3$$

$$f(.5) = (.5)^2 - (.5) + 3 = 2.75$$

$$f(1) = (1)^2 - (1) + 3 = 3$$

$$f(1.5) = (1.5)^2 - (1.5) + 3 = 3.75$$

$$f(2) = (2)^2 - (2) + 3 = 5$$

$$f(2.5) = (2.5)^2 - (2.5) + 3 = 6.75$$

<u>Insert</u> these values into the equation.

$$(\frac{1}{2})[3 + 2.75 + 3 + 3.75 + 5 + 6.75]$$

Simplify:

$$(\frac{1}{2})[24.25]$$

12.125

The accumulated area found of $f(x) = x^2 - x + 3$ a=0   b=3 LH_6 is 12.125. To see what has been done visually, a graph is below. The function $f(x) = x^2 - x + 3$ is the curved line and the total area found (12.125) through six left-hand Riemann sums is shaded area. The area found through this technique is less than the actual area.

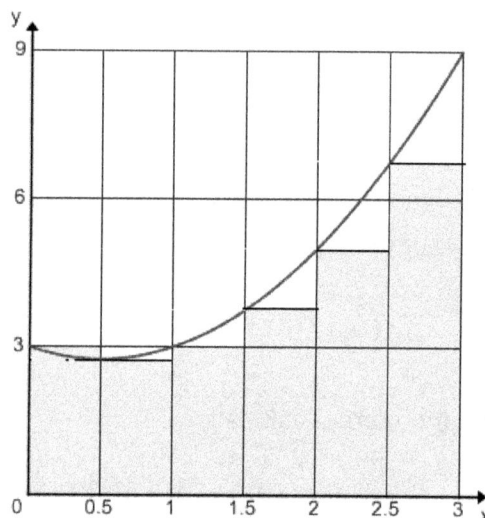

Example 22a-3:
Left-Hand Riemann Sum

Find the area if f(x)=sin(x)   a=0 $b = \pi$ LH_4

Find the value of the intervals (n=4).

$$\frac{b-a}{n}$$

$$\frac{\pi - 0}{4}$$

$\frac{\pi}{4}$ is the interval for the problem.

Set the left hand equation up as (i)[f(a)+f(a+i)+f(a+2i)+f(a+3i)+......] where $\frac{\pi}{4}$ is i.

$$(\frac{\pi}{4})[\sin(0) + \sin(0 + \frac{\pi}{4}) + \sin(0 + \frac{\pi}{2}) + \sin(0 + \frac{3\pi}{4})]$$

Now find the correct values for sin(_)

$$\sin(0) = 0$$

$$\sin(\frac{\pi}{4}) = .707$$

$$\sin(\frac{\pi}{2}) = 1$$

$$\sin(\frac{3\pi}{4}) = .707$$

Insert these values into the equation.

$$(\frac{\pi}{4})[0 + .707 + 1 + .707]$$

Convert $\frac{\pi}{4}$ into a decimal and insert into the equation.

$$\frac{\pi}{4} = .785$$

$$(.785)[0 + .707 + 1 + .707]$$

Simplify

$$(.785)[2.414]$$

$$1.895$$

The accumulated area found of f(x)=sin(x)   a=0 $b = \pi$ LH_4 is 1.895. This can be seen on the graph below with the function as a curved line and accumulated area as a shaded region. With this Riemann sum, the left-hand method accumulated slightly less area than the actual area under the curve.

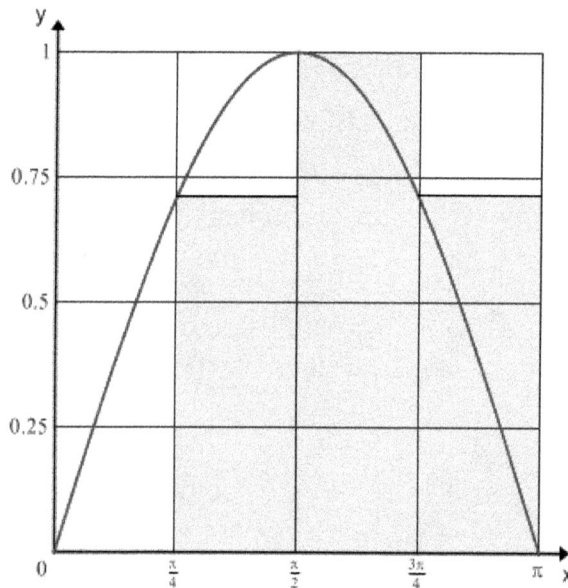

b. Right-Hand Riemann Sums

The difference between left-hand and right-hand Riemann sums is shown below.

Left-hand: (i)[f(a)+f(a+i)+f(a+2i)+f(a+3i)+......]

Right-hand: (i)[f(a+i)+f(a+2i)+f(a+3i)+f(a+4i)+...... +f(b)]

Example 22b-1:
Right Hand Riemann Sum

Find the area if $f(x) = \dfrac{1}{x}$ a=0   b=3 RH_4

The first step to solve this problem is find the value of the intervals. The interval can be found with the equation $\dfrac{b-a}{n}$. N is 4. a and b are given.

$$\frac{b-a}{n}$$

$$\frac{3-1}{4}$$

$\dfrac{1}{2}$ is the interval we will be using for this problem.

58

Insert appropiate information into the RH equation of
(i)[f(a+i)+f(a+2i)+f(a+3i)+f(a+4i)+…… +f(b)] where i= ½. There will be four terms
in total because n is 4.

$$(\frac{1}{2})[f(1.5)+f(2)+f(2.5)+f(3)]$$

Find the correct values for f(_)

$$f(1.5) = \frac{1}{1.5} = \frac{2}{3}$$

$$f(2) = \frac{1}{2}$$

$$f(2.5) = \frac{1}{2.5} = \frac{2}{5}$$

$$f(3) = \frac{1}{3}$$

Substitute these values into the equation.

$$(\frac{1}{2})[\frac{2}{3}+\frac{1}{2}+\frac{2}{5}+\frac{1}{3}]$$

Simplify:

$$(\frac{1}{2})[1.9]$$

.95

The accumulated area found of $f(x) = \frac{1}{x}$,   a=1   b=3 RH_4 is .95. To see

what has been done visually, a graph is below. The function $f(x) = \frac{1}{x}$ is the

curved line and the total area found (.95) through four right-hand Riemann sums
is shaded. Note: .95 is less than the actual area under the function and the LH
area found earlier (1.2835) for the same function.

59

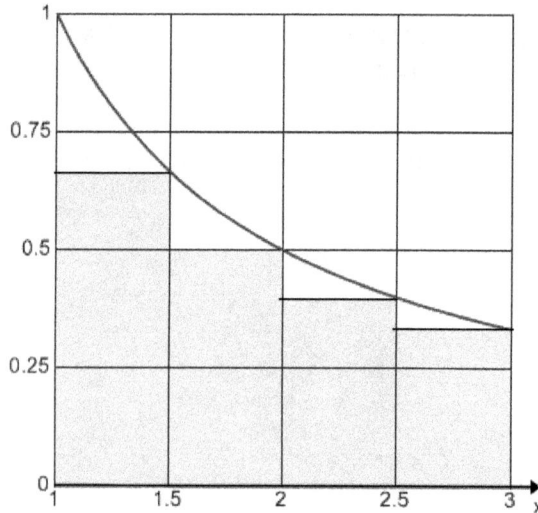

Example 22b-2:
Right-Hand Riemann Sum

Find the area if $f(x) = x^2 - x + 3$ a=0   b=3 RH_6

The first step to solve this question is find the value of the intervals. This can be found by using the equation $\dfrac{b-a}{n}$. N is 6.

$$\dfrac{b-a}{n}$$

$$\dfrac{3-0}{6}$$

$\dfrac{1}{2}$ is the interval we will be using for this problem.

Substitute the appropiate information into the RH equation of
(i)[f(a+i)+f(a+2i)+f(a+3i)+f(a+4i)+...... +f(b)] where i =1/2. There will be 6 f(_)'s.

$$(\tfrac{1}{2})[f(.5) + f(1) + f(1.5) + f(2) + f(2.5) + f(3)]$$

Calculate the correct values for f(_)

$$f(.5) = (.5)^2 - (.5) + 3 = 2.75$$

$$f(1) = (1)^2 - (1) + 3 = 3$$

60

$$f(1.5) = (1.5)^2 - (1.5) + 3 = 3.75$$

$$f(2) = (2)^2 - (2) + 3 = 5$$

$$f(2.5) = (2.5)^2 - (2.5) + 3 = 6.75$$

$$f(3) = (3)^2 - (3) + 3 = 9$$

Insert these values into the equation.

$$(\frac{1}{2})[2.75 + 3 + 3.75 + 5 + 6.75 + 9]$$

Simplify

$$(\frac{1}{2})[30.25]$$

$$15.125$$

The accumulated area found of $f(x) = x^2 - x + 3$ a=0   b=3 RH_6 is 15.125. A graph of this problem is below. The function $f(x) = x^2 - x + 3$ is the curved line and the total area found with six right-hand Riemann sums is shaded. This area (15.125) is greater than the actual area and the area found with left-hand Riemann sums.

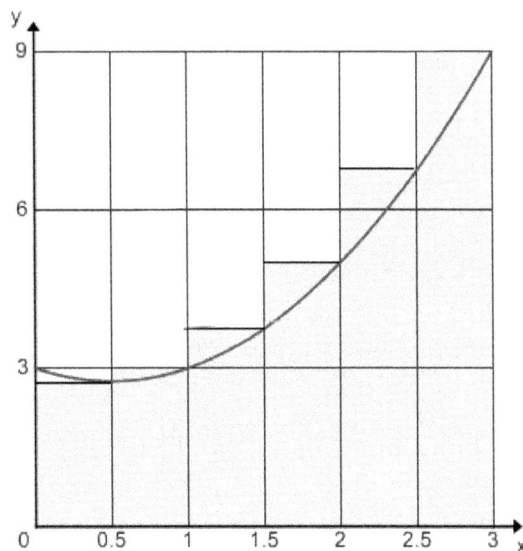

Example 22b-3:
Right-Hand Riemann Sum

Find the area if f(x)=sin(x)   a=0 $b = \pi$ RH_4

Find the value of the intervals (n=4)

$$\frac{b-a}{n}$$

$$\frac{\pi-0}{4}$$

$\dfrac{\pi}{4}$ is the interval for the problem.

List the right-hand equation as (i)[f(a+i)+f(a+2i)+f(a+3i)+f(a+4i)+...... +f(b)] where $\dfrac{\pi}{4}$ is i. There will be four values we need to find.

$$(\frac{\pi}{4})[\sin(0+\frac{\pi}{4})+\sin(0+\frac{\pi}{2})+\sin(0+\frac{3\pi}{4})+\sin(\pi)]$$

Next, find the correct values for sin(_)

$$\sin(\frac{\pi}{4}) = .707$$

$$\sin(\frac{\pi}{2}) = 1$$

$$\sin(\frac{3\pi}{4}) = .707$$

$$\sin(\pi) = 0$$

Substitute these values into the equation.

$$(\frac{\pi}{4})[.707+1+.707+0]$$

Convert $\dfrac{\pi}{4}$ into a decimal and substitute into the equation.

$$\frac{\pi}{4} = .785$$

$$(.785)[.707 + 1 + .707 + 0]$$

Simplify

$$(.785)[2.414]$$

1.895

The accumulated area found of f(x)=sin(x)   a=0 $b = \pi$ RH_4 is 1.895. With this right-hand Riemann sum method, as with the left-hand method, less area was accumulated than what actually is under the function.

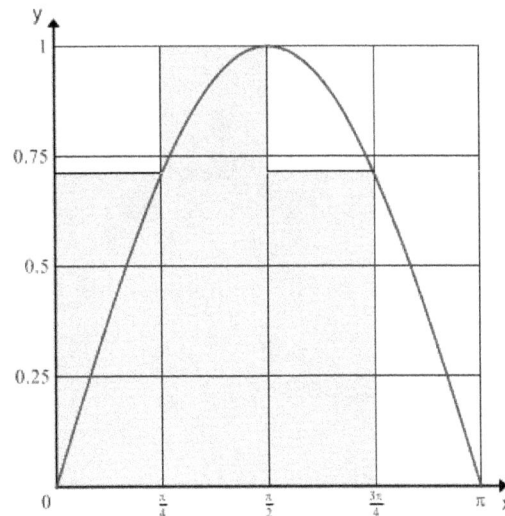

c. Midpoint Riemann Sums

The only difference between left-hand, right-hand, and midpoint Riemann sums is that you calculate the midpoint of the interval (.5i) and add it to a. All three methods can be seen below. Midpoint Riemann sums tend to be slightly more accurate than one found through the right- or left-hand methods.

Left hand: (i)[f(a)+f(a+i)+f(a+2i)+f(a+3i)+......]

Right hand: (i)[f(a+i)+f(a+2i)+f(a+3i)+f(a+4i)+...... +f(b)]

Midpoint: (i)[f(a+.5i)+f(a+1.5i)+f(a+2.5i)+f(a+3.5i)+......]

Example 22c-1:
Midpoint Riemann Sum

Find the area if $f(x) = \dfrac{1}{x}$ a=0   b=3 M_4

The first step to solve this problem is find the value of the intervals. It can be found with the equation $\dfrac{b-a}{n}$. n=4

$$\frac{b-a}{n}$$

$$\frac{3-1}{4}$$

$\dfrac{1}{2}$ is the interval we will be using for this problem.

Use the midpoint equation of (i)[f(a+.5i)+f(a+1.5i)+f(a+2.5i)+f(a+3.5i)+......] where i= ½. There will be four terms that need to be found.

$$(\frac{1}{2})[f(1.25)+f(1.75)+f(2.25)+f(2.75)]$$

Now find the correct values for f(_)

$$f(1.25) = \frac{1}{1.25} = \frac{4}{5}$$

$$f(1.75) = \frac{1}{1.75} = \frac{4}{7}$$

$$f(2.25) = \frac{1}{2.25} = \frac{4}{9}$$

$$f(2.75) = \frac{1}{2.75} = \frac{4}{11}$$

Insert these values into the equation.

$$(\frac{1}{2})[\frac{4}{5}+\frac{4}{7}+\frac{4}{9}+\frac{4}{11}]$$

64

$$(\frac{1}{2})[2.180]$$

1.09

The accumulated area of $f(x) = \frac{1}{x}$, a=1   b=3 M_4 is 1.09. The function

$f(x) = \frac{1}{x}$ is the curved line and the total area found (1.09) through four midpoint

Riemann sums is shaded. The total area found for the same function with the midpoint method is less than the LH area found earlier (1.2835) but more than the RH area (.95).

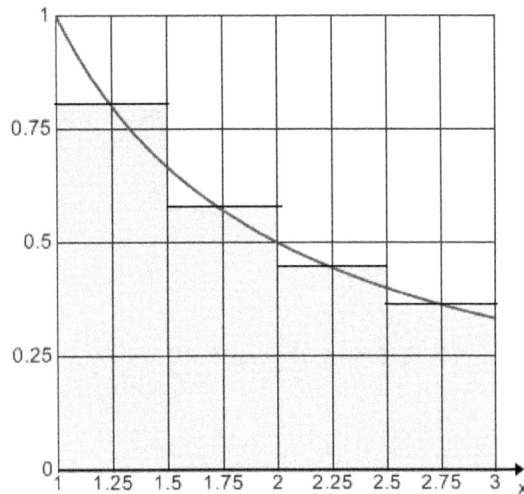

Example 22c-2:
Midpoint Riemann Sum

Find the area if $f(x) = x^2 - x + 3$ a=0   b=3 M_6

The first step to solve this problem is find the value of the intervals. They can be

found by using the equation $\dfrac{b-a}{n}$. n=6

$$\frac{b-a}{n}$$

$$\frac{3-0}{6}$$

$\frac{1}{2}$ is the interval we will be using for this problem.

Substitute appropiate information into the midpoint equation of
(i)[f(a+.5i)+f(a+1.5i)+f(a+2.5i)+f(a+3.5i)+......] where i =1/2. There will be six
terms (f(_)'s) total.

$$(\frac{1}{2})[f(.25) + f(.75) + f(1.25) + f(1.75) + f(2.25) + f(2.75)]$$

Find the correct values for f(_)

$$f(.25) = (.25)^2 - (.25) + 3 = 2.8125$$

$$f(.75) = (.75)^2 - (.75) + 3 = 2.8125$$

$$f(1.25) = (1.25)^2 - (1.25) + 3 = 3.3125$$

$$f(1.75) = (1.75)^2 - (1.75) + 3 = 4.3125$$

$$f(2.25) = (2.25)^2 - (2.25) + 3 = 5.8125$$

$$f(2.75) = (2.75)^2 - (2.75) + 3 = 7.8125$$

Replace f(_) with the values found above.

$$(\frac{1}{2})[2.8125 + 2.8125 + 3.3125 + 4.3125 + 5.8125 + 7.8125]$$

Simplify

$$(\frac{1}{2})[26.875]$$

$$13.4375$$

The accumulated area found of $f(x) = x^2 - x + 3$ a=0   b=3 M_6 is
13.4375. The function $f(x) = x^2 - x + 3$ is the curved line and the total area
found (13.4375) through six midpoint Riemann sums is the grey shaded region.

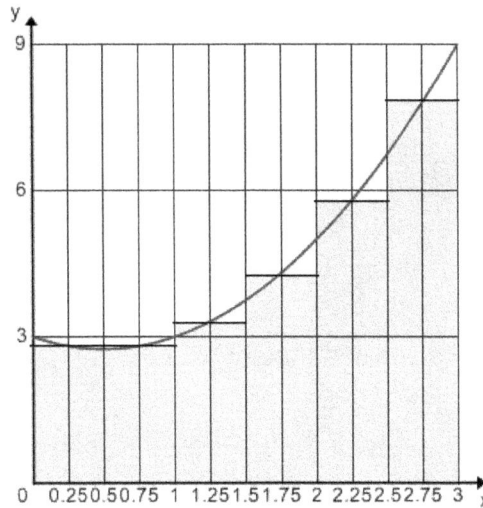

Example 22c-3:
Midpoint Riemann Sum

Find the area if f(x)=sin(x)   a=0 $b = \pi$ M_4

Find the value of the intervals (n=4)

$$\frac{b-a}{n}$$

$$\frac{\pi - 0}{4}$$

$\frac{\pi}{4}$ is the interval for the problem.

Remember to set the midpoint equation up as

(i)[f(a+.5i)+f(a+1.5i)+f(a+2.5i)+f(a+3.5i)+......] where $\frac{\pi}{4}$ is i. There will be four

terms.

$$(\frac{\pi}{4})[\sin(0+\frac{\pi}{8})+\sin(0+\frac{3\pi}{8})+\sin(0+\frac{5\pi}{8})+\sin(0+\frac{7\pi}{8})]$$

Calculate the values for sin(_)

$$\sin(\frac{\pi}{8}) = .383$$

$$\sin(\frac{3\pi}{8}) = .924$$

$$\sin(\frac{5\pi}{8}) = .924$$

$$\sin(\frac{7\pi}{8}) = .383$$

Insert these values into the equation.

$$(\frac{\pi}{4})[.383 + .924 + .924 + .383]$$

Convert $\frac{\pi}{4}$ into a decimal and substitute into the equation.

$$\frac{\pi}{4} = .785$$

$$(.785)[.383 + .924 + .924 + .383]$$

Simplify

$$(.785)[2.614]$$

$$2.052$$

The accumulated area found of f(x)=sin(x) a=0 $b = \pi$ M_4 is 2.052. This can be seen on the graph below where the original function is the curved line and accumulated area is shaded. The midpoint method for this problem accounted for more area than is actually under the curve.

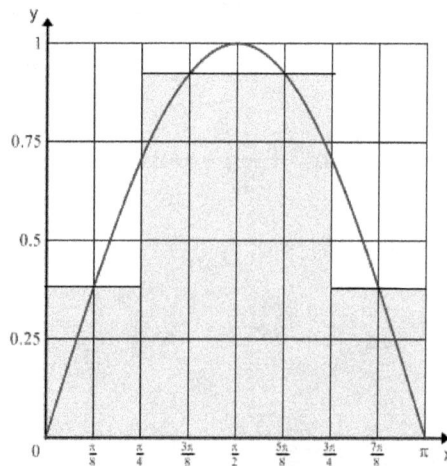

23. Trapezoid Rule

The trapezoid rule is similar to the right hand, left hand, and midpoint riemann sums but instead of finding areas in the form rectangles, it does so with trapezoids.

Left hand: (i)[f(a)+f(a+i)+f(a+2i)+f(a+3i)+......]

Right hand: (i)[f(a+i)+f(a+2i)+f(a+3i)+f(a+4i)+......+f(b)]

Midpoint: (i)[f(a+.5i)+f(a+1.5i)+f(a+2.5i)+f(a+3.5i)+......]

Trapezoid rule: (.5i)[(f(a)+f(a+i))+(f(a+i)+f(a+2i))+(f(a+2i)+f(a+3i))+....+(f(a+_i)+f(b))]

Example 23:
Trapezoid Rule

Find the area if $f(x) = x^2 - x + 3$ a=0   b=3 T_6

The first step to solve this problem, just like the other Riemann sums, is to find the value of the intervals. The interval value can be found by using the equation $\dfrac{b-a}{n}$.

$$\frac{b-a}{n}$$

$$\frac{3-0}{6}$$

$\dfrac{1}{2}$ is the interval.

Insert the appropiate information into the trapezoid equation of
(.5i)[(f(a)+f(a+i))+(f(a+i)+f(a+2i))+(f(a+2i)+f(a+3i))+....+(f(a+_i)+f(b))] where i =1/2

$$((\frac{1}{2})(\frac{1}{2}))[(f(0)+f(.5))+(f(.5)+f(1))+(f(1)+f(1.5))+(f(1.5)+f(2))+$$
$$(f(2)+f(2.5))+(f(2.5)+f(3))]$$

Now find the correct values for f(_)

$$f(0) = (0)^2 - (0) + 3 = 3$$

$$f(.5) = (.5)^2 - (.5) + 3 = 2.75$$

$$f(1) = (1)^2 - (1) + 3 = 3$$

69

$$f(1.5) = (1.5)^2 - (1.5) + 3 = 3.75$$

$$f(2) = (2)^2 - (2) + 3 = 5$$

$$f(2.5) = (2.5)^2 - (2.5) + 3 = 6.75$$

$$f(3) = (3)^2 - (3) + 3 = 9$$

Insert these values into the equation.

$$((\frac{1}{2})(\frac{1}{2}))[(3 + 2.75) + (2.75 + 3) + (3 + 3.75) + (3.75 + 5) +$$
$$(5 + 6.75) + (6.75 + 9)]$$

Simplify

$$(\frac{1}{4})[(3 + 2.75) + (2.75 + 3) + (3 + 3.75) + (3.75 + 5) +$$
$$(5 + 6.75) + (6.75 + 9)]$$

$$(\frac{1}{4})(54.5)$$

13.625

The accumulated area found of $f(x) = x^2 - x + 3$ a=0   b=3 T_6 is 13.625.

The function $f(x) = x^2 - x + 3$ is the curved line and the total area found (13.625) through six trapezoid Riemann sums is shaded on the graph below. Notice how the connecting lines between each interval almost entirely overlap with the actual function. The trapezoid rule of Riemann sums usually provides the most accurate results compared to right hand, left hand, and midpoint methods. However, there are easier ways to calculate the exact area under a curve.

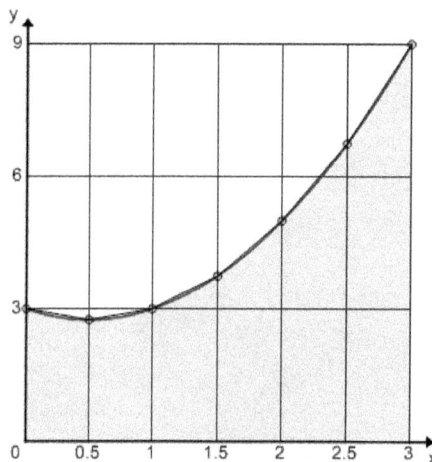

70

24. How to Find Definite Integrals on the TI-83 Graphing Calculator

Procedure:

Open "Y="

Plot function in " $y_1 =$ "

Graph the function

Go into "2nd mode" and push "Trace" button (also known as "Calc" button in 2nd mode)

Select the 7th option which is " $\int f(x)dx$ "

Enter the lower limit x value ($\int_x f(x)$) and press enter

Enter the upper limit x value ($\int^x f(x)$)

Push "Enter" and the area accumulated should appear in the lower left hand corner.

Try it yourself with the following example: $\int_2^6 (7x^2 - 13x)dx$

The correct answer is 277.333.

Using the TI-83 or other graphing calculator will save time and will allow you to check your work. However, the steps required to find integrals must be known for situations where you cannot use a calculator.

25. The Fundamental Theorem of Calculus

Part I

If f(x) is a continuous function and the integral (can also be referred to as area accumulating function or antiderivative) of function is f(x)= $\int_a^x f(t)$ then the derivative of the area accumulating function equals f(x) OR $\frac{dy}{dx} \int_a^x f(t)dt$ equals f(x).

71

Part II

$\int_a^b f(x)dx = F(b) - F(a)$ where F(x) is an antiderivative of f(x)

Example 25a-1:
Part 1 of the Fundamental Theorem of Calculus

$$\int_0^x \frac{\sin(t)}{t}dt \quad \rightarrow \quad \frac{d}{dx}[\int_0^x \frac{\sin(t)}{t}dt] = \frac{\sin(x)}{x}$$

Example 25a-2:
Part 1 of the Fundamental Theorem of Calculus

$$A_{e^x} \rightarrow \frac{d}{dx}[A_{e^x}] = e^x$$

Example 25b-1:
Part 2 of the Fundamental Theorem of Calculus

Prove $\int_0^1 2x^2 dx = \frac{2}{3}x^3\Big|_0^1$

<u>Find</u> what each side of the equation is equal to. It should be the same value.

$$\int_0^1 2x^2 dx = \frac{2}{3} \text{ (found with TI-83)}$$

Now <u>find</u> what $\frac{2}{3}x^3\Big|_0^1$ is. It should be 2/3.

$$\frac{2}{3}x^3\Big|_0^1 = \frac{2}{3}(1)^3 - \frac{2}{3}(0)^3$$

$$\frac{2}{3}x^3\Big|_0^1 = \frac{2}{3}$$

Both sides are equal as stated in part two of the fundamental theorem.

Example 25b-2:
Part 2 of the Fundamental Theorem of Calculus

$$\int_{5}^{9} 3x + 2 = \frac{3}{2}x^2 + 2x \Big|_{5}^{9}$$

<u>Find</u> what each side of the equation is equal to. It should be the same value.

$$\int_{5}^{9} 3x + 2 = 92 \text{ (found with TI-83)}$$

Now <u>find</u> what $\frac{3}{2}x^2 + 2x \Big|_{5}^{9}$ is. It should be 92.

$$\frac{3}{2}x^2 + 2x \Big|_{5}^{9} = ((\frac{3}{2})(9)^2 + 2(9)) - ((\frac{3}{2})(5)^2 + 2(5))$$

$$\frac{3}{2}x^2 + 2x \Big|_{5}^{9} = (139.5) - (47.5)$$

$$\frac{3}{2}x^2 + 2x \Big|_{5}^{9} = 92$$

Both sides are equal (see part two of the fundamental theorem).

26. *Integration Techniques*

Before we cover specific integration techniques, first we need to establish the difference between definite and indefinite integrals.

Indefinite integral= $\int f(x)dx$

The indefinite integral does not have endpoints.

Definite integral= $\int_{a}^{b} f(x)dx$

The definite integral finds the area accumulated between the endpoints a and b.

a. The Power Rule of Integration

The power rule of integration is $\int x^k dx = (\frac{1}{k+1})(x^{k+1})$, $k \neq -1$

Example 26a:
The Power Rule of Integration

Integrate $\int \dfrac{(x-4)^2}{\sqrt{x}} dx$

First <u>multiply</u> out (x-4)^2

$$\int \dfrac{x^2 - 8x + 16}{\sqrt{x}} dx$$

<u>Divide</u> each term of the numerator by $\sqrt{x}$

$$\int x^{\frac{3}{2}} - 8x^{\frac{1}{2}} + 16x^{\frac{-1}{2}} dx$$

Now <u>apply</u> the power rule to each term

$$\int x^{\frac{3}{2}} dx = (\dfrac{1}{\frac{3}{2}+1})(x^{\frac{3}{2}+1})$$

$$\int x^{\frac{3}{2}} dx = \dfrac{2}{5} x^{\frac{5}{2}}$$

$$\int -8x^{\frac{1}{2}} dx = (-8)(\dfrac{1}{\frac{1}{2}+1})(x^{\frac{1}{2}+1})$$

$$\int -8x^{\frac{1}{2}} dx = -\dfrac{16}{3} x^{\frac{3}{2}}$$

$$\int 16x^{\frac{-1}{2}} dx \quad \begin{aligned} = (16)(\dfrac{1}{\frac{-1}{2}+1})(x^{\frac{-1}{2}+1}) \end{aligned}$$

$$\int 16x^{\frac{-1}{2}} dx = 32x^{\frac{1}{2}}$$

Next <u>sum</u> all these individual terms together and <u>add</u> the +C.

74

$$\frac{2}{5}x^{\frac{5}{2}} - \frac{16}{3}x^{\frac{3}{2}} + 32x^{\frac{1}{2}} + C$$

Note: +C is the constant of integration and must be added to all indefinite integrals.

The integral of $\int \frac{(x-4)^2}{\sqrt{x}}dx$ is $\frac{2}{5}x^{\frac{5}{2}} - \frac{16}{3}x^{\frac{3}{2}} + 32x^{\frac{1}{2}} + C$.

b. Exponential Integrals

The rule utilized to find exponential integrals is $\int k^{cx}dx = (\frac{1}{(c)(\ln k)})(k^{cx})$

Example 26b:
Exponential Integrals

Integrate $\int 7^x dx$

<u>Use</u> the equation $\int k^{cx}dx = (\frac{1}{(c)(\ln k)})(k^{cx})$ and <u>substitute</u>.

$$\int 7^x dx = (\frac{1}{(1)(\ln 7)})(7^x)$$

<u>Simplify</u> and <u>add</u> +C since we are finding an indefinite integral

$$\int 7^x dx = (\frac{7^x}{\ln 7}) + C$$

c. "E" Integrals

The following rule is used where the mathematical constant "e" is present in the function

$$\int e^{cx}dx = \frac{1}{c}e^{cx}$$

Example 26c:
"E" Integrals:

Integrate $\int e^{-4x}dx$

$$\int e^{-4x}dx = \frac{1}{-4}e^{-4x}$$

<u>Add</u> the +C

$$\int e^{-4x}dx = -\frac{1}{4}e^{-4x} + C$$

d. Integrals with Logarithms

Use $\int \frac{1}{ax+b}dx = \frac{1}{a}\ln|ax+b|$ when integrating a 1/x function.

Example 26d:
Integrals with Logarithms

Integrate $\int \frac{1}{2x}dx$

$$\int \frac{1}{2x}dx = \frac{1}{2}\ln|2x|$$

<u>Add</u> the +C

$$\int \frac{1}{2x}dx = \frac{1}{2}\ln|2x| + C$$

e. Common Trigonometric Integrals

$\int \sin(x)dx$	$= -\cos(x) + C$
$\int \cos(x)dx$	$= \sin(x) + C$
$\int \cot(x)dx$	$= \ln|\sin(x)| + C$
$\int \tan(x)dx$	$= \ln|\sec(x)| + C$
$\int \sec^2(x)dx$	$= \tan(x) + C$
$\int \csc^2(x)dx$	$= -\cot(x) + C$

27. Integration by Parts and Substitution
a. Integration by Parts

Some integrals cannot be found with the techniques discussed in the previous section. Integration by parts and integration by substitution will often solve these difficult integrals.

Integration by parts is best when integrating two functions that are multiplied together. The formula utilized for integration by parts is $\int u\,dv = uv - \int v\,du$

Example 27a:
Integration by Parts

Integrate $\int x\cos(4x)dx$

In this problem you need to choose what will be u and what will be dv of the given equation. Let u=x and dv=cos(4x).

Find du and v

du is the derivative of u. u is x. The derivative of x is (1)dx or just dx.

v is the antiderivative of dv. dv is cos(4x). The antiderivative of cos(4x) is (1/4)sin(4x).

Now insert this information into $\int u\,dv = uv - \int v\,du$

$$\int x\cos(4x)dx = (x)(\frac{1}{4})\sin(4x) - \int(\frac{1}{4})\sin(4x)dx$$

Simplify

$$(\frac{1}{4})x\sin(4x) - (\frac{1}{4})\int\sin(4x)dx$$

Integrate $\int(\frac{1}{4})\sin(4x)dx$

$$\int(\frac{1}{4})\sin(4x)dx = \frac{-\cos(4x)}{4} + C$$

Insert this information back into the parts equation.

$$(\frac{1}{4})x\sin(4x) - (\frac{1}{4})(\frac{-\cos(4x)}{4}) + C$$

Simplify

77

$$(\frac{1}{4})x\sin(4x)+(\frac{\cos(4x)}{16})+C$$

This is the final answer:

$$\int x\cos(4x)dx=(\frac{1}{4})x\sin(4x)+(\frac{\cos(4x)}{16})+C$$

b. Integration by Substitution

Integration by substitution is the counterpart to the chain rule of differentiation. Use the terms u and u's derivative, du, to find the integral.

Example 27b:
Integration by Substitution

Integrate $\int \dfrac{4x+8}{x^2+4x+1}dx$

The first step is to <u>select</u> u. In this case the denominator of x^2+4x+1 is the best option.

<u>Find</u> du. To find du, differentiate u.

du=(2x+4)dx

<u>Substitute</u> u and du into the equation. The numerator is twice the size of du so it is 2du.

$$\int \frac{2du}{u}$$

<u>Simplify</u>

$$2\int \frac{1}{u}du$$

<u>Integrate</u> (without substituting)

$$2\ln|u|+C$$

<u>Substitute</u> x^2+4x+1 for u

$$2\ln\left|x^2 + 4x + 1\right| + C$$

$$\int \frac{4x + 8}{x^2 + 4x + 1}\,dx = 2\ln\left|x^2 + 4x + 1\right| + C$$

Part V: Applications of Integration

28. Area Accumulation Functions

Given y=f(x) which is graphed below.

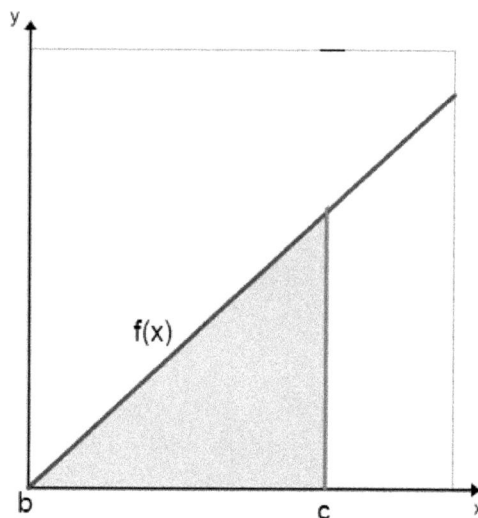

In order to find the shaded area (a definite integral) <u>use</u> the formula $A_f(x) = \int_b^c f(x)dx$.

This area accumulation function sums all the area between two x-values (b and c in this case) with the integral of the function.

Example 28a:
Area Accumulation Functions

<u>Find</u> the shaded area under the curve.

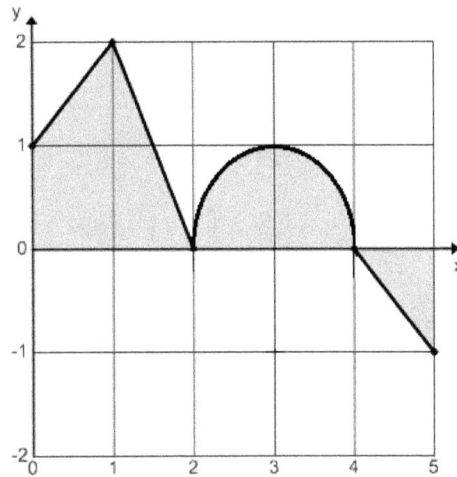

Sometimes the area has to be found without a function to work with. <u>Break</u> each segment into sections which the area can be found.

Between 0<x<1 there are 1.5 units. A 1x1 square and a .5 unit triangle (.5 times a 1 unit base times a 1 unit height).

Between 1<x<2 there is 1 unit triangle (.5 times a 1 unit base times a 2 unit height).

Between 2<x<4 there are $\frac{1}{2}\pi$ units. (The area is a semicircle and the formula to find a semicircle's area is $\frac{\pi}{2}r^2$. The radius of the semicircle is 1 so the area is $\frac{1}{2}\pi$ units.)

Between 4<x<5 there is a -.5 unit triangle (.5 times a 1 unit base times a 1 unit height). This area is negative because it is below the x-axis.

<u>Sum</u> all the areas found above.

$$1.5+1+\frac{1}{2}\pi-.5$$

The final answer is $\int_0^5 f(x)dx = 2+\frac{1}{2}\pi$

Example 28b-1:
Area Accumulation Functions

Given the following: $\int_0^2 f(x)dx = 5$, $\int_2^4 f(x)dx = -2$, and $\int_0^6 f(x)dx = 6$.

Find $\int_2^6 3f(x)dx$

For this problem you need to <u>multiply</u> the difference between integrals by three

$$3(\int_0^6 f(x)dx - \int_0^2 f(x)dx)$$

$\int_0^6 f(x)dx - \int_0^2 f(x)dx$ are already given above so <u>substitute</u> and <u>solve</u>.

3(6-5)

3(1)

$$\int_2^6 3f(x)dx = 3$$

Example 28b-2:
Area Accumulation Functions

Given the same, $\int_0^2 f(x)dx = 5$, $\int_2^4 f(x)dx = -2$, and $\int_0^6 f(x)dx = 6$:

Find $\int_0^2 f(x) + 5dx$

With this problem, $\int_0^2 f(x)$ is already known to be 5.

$$5 + \int_0^2 5dx$$

Next <u>find</u> the area under y=5 from x=0 to x=2

5(2)-5(0)

10

<u>Carry down</u> the 5 from the first part of the question

5+10

$$\int_0^2 f(x) + 5dx = 15$$

29. Position, Velocity, and Acceleration

Position----(differentiate)---- $\rightarrow$ Velocity----(differentiate)---- $\rightarrow$ Acceleration

The absolute value of velocity is speed.

Acceleration----(integrate)---- $\rightarrow$ Velocity----(integrate)---- $\rightarrow$ Position

$|$Velocity$|$ ----(integrate)---- $\rightarrow$ Total distance travelled

Example 29a:
Velocity and Position

The acceleration of a particle moving from left to right is $a(t) = t^2 - 4 \dfrac{ft}{second^2}$.

$v(1) = \dfrac{4}{3} \dfrac{ft}{second}$ and its postion after 1 second was $\dfrac{13}{12}$ feet. Find v(t) and p(t).

The first step is to <u>find</u> the integral of a(t)=t^2-4. The power rule is the best method.

$$\int t^2 - 4 \, dx = \frac{1}{3}t^3 - 4t + C$$

$$v(t) = \frac{1}{3}t^3 - 4t + C$$

Now <u>find</u> +C. We know v(1)=4/3 so <u>substitute</u> the appropiate values to find C

$$\frac{4}{3} = \frac{1}{3}(1)^3 - 4(1) + C$$

$$\frac{4}{3} = \frac{1}{3} - 4 + C$$

$$1 = -4 + C$$

$$C = 5$$

$$v(t) = \frac{1}{3}t^3 - 4t + 5$$

Now to find p(t), <u>find</u> the integral of v(t).

83

$$\int \frac{1}{3}t^3 - 4t + 5 dx = \frac{1}{12}t^4 - 2t^2 + 5t + C$$

$$p(t) = \frac{1}{12}t^4 - 2t^2 + 5t + C$$

Once again +C must be found. Fortunately, p(1) is known to be 13/12.

$$\frac{13}{12} = \frac{1}{12}(1)^4 - 2(1)^2 + 5(1) + C$$

$$\frac{13}{12} = \frac{1}{12} - 2 + 5 + C$$

$$1 = -2 + 5 + C$$

$$-4 = -2 + C$$

$$-2 = C$$

$$p(t) = \frac{1}{12}t^4 - 2t^2 + 5t - 2$$

Example 29b:
Velocity and Acceleration

The position of an atom is represented by the function $p(t) = \cos(2t) + e^{2t} + 3t^7$

Find v(t) and a(t).

For this problem we need to differentiate, not integrate like the last example, to find v(t) and a(t).

$$p(t) = \cos(2t) + e^{2t} + 3t^7$$

$$p'(t) = v(t)$$

$$p'(t) = -2\sin(2t) + 2e^{2t} + 21t^6$$

$$v(t) = -2\sin(2t) + 2e^{2t} + 21t^6$$

The process is the same to find a(t) but replace p(t) with v(t)

$$v(t) = -2\sin(2t) + 2e^{2t} + 21t^6$$

$$v'(t) = a(t)$$

$$v'(t) = -4\cos(2t) + 4e^{2t} + 126\,x^5$$

$$a(t) = -4\cos(2t) + 4e^{2t} + 126\,x^5$$

30. Net Distance and Total Distance

Net distance refers to how far something has moved from an origin point. When an object changes direction, the net distance will no longer equal total distance. Total distance solely focuses on how far the object has traversed regardless of direction.

The best way to tell the two apart is with a real life example. Assume the origin is your residence. Net distance (also sometimes called "displacement") is how far you are from your home at any time (often zero!). The total distance would be like your personal odometer which includes trips to the grocery store, work, school, etc that is always increasing (and thus positive).

Below are the equations required to calculate net and total distance. An integral is needed since we are trying to find position with a velocity equation.

Net distance= $\int_b^a f(x)\,dx$

Total distance= $\int_b^a \left| f(x) \right| dx$

Example 30a:
Net and Total Distance
(Find with TI-83; refer to page 65 if necessary)

Find the net and total distance traveled from .25 seconds to six seconds if the equation is $v(t) = \dfrac{-\cos(6t)}{t}$

Net distance= $\int_{.25}^{6} \dfrac{-\cos(6t)}{t}\,dt = $ .4978 units to the right of the origin

Total distance= $\int_{.25}^{6} \left| \dfrac{-\cos(6t)}{t} \right| dt = 1.9633$ units traveled

Example 30b:
Net and Total Distance

Find the net and total distance traversed in the time frame $8 \geq t \geq 0$ if v(t)= $1.2t^2 - 5$

Net distance= $\int_0^8 (1.2t^2 - 5)dt$

Follow the power rule of integration

$$.4t^3 - 5t \Big|_0^8$$

First <u>find</u> t(8) then <u>subtract</u> t(0) from that total

$$t(8)= .4(8^3) - 5(8)$$

$$.4(512)\text{-}40$$

$$204.8\text{-}40= 164.8$$

$$t(0) = .4(0^3) - 5(0\_)$$

$$t(0)=0$$

$$164.8\text{-}0=164.8$$

The net distance from the origin is 164.8 units

Total distance $= \int_0^8 |1.2t^2 - 5|dt$

Since this is an absolute value function we need to <u>check</u> for zeros

$$1.2t^2 - 5 = 0$$

$$1.2t^2 = 5$$

$$t^2 = 4.1667$$

$$t = 2.041$$

Now we can utilize the integral power rule but we split it up with $\left|.4t^3 - 5t\right|_0^{2.041}$ and $\left|.4t^3 - 5t\right|_{2.041}^8$. Once we find one that is negative, we <u>double</u> that amount and <u>add</u> it to the net distance amount of 164.8 units.

86

$$\left| .4t^3 - 5t \right|_0^{2.041}$$

$$\left| .4(2.041^3) - 5(2.041) \right|$$

$$\left| .4(8.502) - 10.205 \right|$$

$$\left| 3.401 - 10.205 \right|$$

$$\left| -6.804 \right|$$

This becomes positive because we are finding the absolute value.

$$\left| .4(0^3) - 5(0) \right| = 0$$

$$\left| .4t^3 - 5t \right|_0^{2.041} = 6.804$$

In the net distance question this 6.084 units was subtracted from the integral since it was "under the curve". Another way to think of it was the the majority of the movement was to the right of the origin but at one point the particle moved 6.084 units to the left before reversing direction. The particle traveled 12.168 total units despite net distance being unchanged.

Add 12.168 to net distance (164.8)

$$12.168 + 164.8 = 176.968 \text{ units}$$

The total distance the particle traversed was 176.968 units.

31. $\dfrac{dy}{dt} = ky$ Separation

Solve for y(t) when $\dfrac{dy}{dt} = ky$ and k is a nonzero constant.

The first step is to separate the variables.

$$\frac{dy}{y} = kdt$$

$$\int \frac{dy}{y} = \int kdt$$

<u>Integrate</u>

$$\ln(y) = kt + C$$

To solve for y, <u>raise</u> both side to the constant e.

$$y(t) = e^{kt+C}$$

$$y(t) = (e^c)(e^{kt})$$

e^c is a constant that we will <u>call</u> y^0

$$y(t) = y_0 e^{kt}$$

This is the same formula as the exponential growth or decay model. Y(0) represents the initial value for y (which is a constant).

32. Separation of Variables

Separation of variables is needed whenever there is an equation with two variables on the same side of the equation.

Example 32:
Separation of Variables

Given $\dfrac{dy}{dx} = 2x(1+y^2)e^{x^2}$ where y(0)=1

<u>Reorganize</u> the problem with all the "x's" on the right side and all the "y's" on the left

$$\frac{dy}{1+y^2} = 2x(e^{x^2})dx$$

Both sides will need to be <u>integrated</u>

$$\int \frac{dy}{1+y^2} = \int 2x(e^{x^2})dx$$

First <u>integrate</u> the "x" side of the equation. Integration by substitution is needed where u= x^2 and du= 2xdx

$$\int 2x(e^{x^2})dx = \int e^u du$$

88

Integrate

$$e^u + C$$

Substitute x^2 for u

$$e^{x^2} + C$$

Now <u>integrate</u> the "y" side

$$\int \frac{dy}{1+y^2} = \arctan(y)$$

$$\arctan(y) = e^{x^2} + C$$

Since we know that y(0)=1, we can <u>substitute</u> these values for x and y to find the constant, C. Note: arctan can also be notated as tan^-1(___).

$$\tan^{-1}(1) = e^0 + C$$

$$\frac{\pi}{4} = 1 + C$$

$$C = \frac{\pi}{4} - 1$$

<u>Substitute</u> $C = \frac{\pi}{4} - 1$

$$\tan^{-1}(y) = (e^{x^2} + \frac{\pi}{4} - 1)$$

<u>Move</u> arctan to the right side of the equation so y is by itself.

$$y = \tan(e^{x^2} + \frac{\pi}{4} - 1)$$

33. Slope Field Example

The graph below is the slope field of $\frac{dy}{dx} = x - .5y$. The slopes of the numerous lines are calculated by inserting the x and y values of each point into the equation. For example, the line at (-1,-2) is horizontal since -1 minus (.5 times -2) is zero. Whereas (-2,0)'s line has a slope of -2.

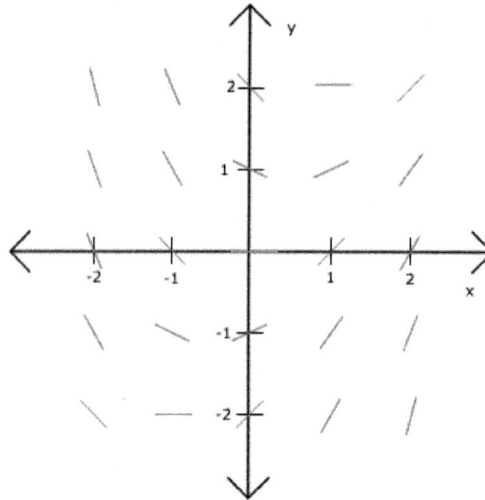

34. Euler's Method

Euler's Method assumes the following:

- A differential equation is given

- An initial value is given

- The size of each margin between estimates or "steps"

This method does not provide a solution of the curve. However, estimated points on the curve are found.

The logic behind Euler's Method is shown below

$$\frac{dy}{dx} = y + x$$

This cannot be solved. However, this equation can be re-written.

$$y_1 - y_0 = m(x_1 - x_0)$$

$$y_1 = y_0 + m(x_1 - x_0)$$

$(x_1 - x_0)$ is the same as dx

$$y_1 = y_0 + m(dx)$$

m is the differential equation (y+x in this case) evaluated at the previous (or given) point.

Substitute the information given to find y_1

Example 34:
Euler's Method

Given $y(1) = 1$ and a step of ¼, <u>estimate</u> y(1.25) and y(1.50) if dy/dx=xy

$$(x_0, y_0) \rightarrow (1,1)$$

$$(x_1, y_1) \rightarrow (1.25, ?)$$

$$(x_2, y_2) \rightarrow (1.5, ?)$$

We must <u>find</u> y(1.25) before being able to find y(1.5). <u>Use</u> the equation
$y = y_0 + m(dx)$

$$y_1 = y_0 + m(dx)$$

y_0 is 1 as indicated in $(x_0, y_0) \rightarrow (1,1)$

$$y_1 = 1 + m(dx)$$

dx is the difference between x_1 (this is 1.25) and x_0 (1). dx is .25

$$y_1 = 1 + m(.25)$$

m is the differential equation (xy) evauluated at the previous point known. That is (1,1) and the product of 1x1=1

$$y_1 = 1 + 1(.25)$$

$$y_1 = 1.25$$

<u>Insert</u> this value into our chart

$$(x_0, y_0) \rightarrow (1,1)$$

$$(x_1, y_1) \rightarrow (1.25, 1.25)$$

$$(x_2, y_2) \rightarrow (1.5, ?)$$

In order to repeat the process for finding y_2, now use $(x_1, y_1) \rightarrow (1.25, 1.25)$ as your known point.

$$y_2 = y_1 + m(dx)$$

y_1 is 1.25 as indicated in $(x_1, y_1) \rightarrow (1.25, 1.25)$

$$y_2 = 1.25 + m(dx)$$

dx is the difference between x_2 (this is 1.5) and x_1 (1.25). dx is .25

$$y_2 = 1.25 + m(.25)$$

m is the differential equation (xy) evaulated at the previous point known. That is (1.25,1.25) and the product of 1.25x1.25=1.5625

$$y_2 = 1.25 + 1.5625\,(.25)$$

$$y_2 = 1.25 + .3906$$

$$y_2 = 1.6406$$

<u>Insert</u> this value into our chart

$$(x_0, y_0) \rightarrow (1, 1)$$

$$(x_1, y_1) \rightarrow (1.25, 1.25)$$

$$(x_2, y_2) \rightarrow (1.5, 1.6406)$$

By utilizing Euler's Method we have been able to find estimations for y_1 and y_2

35. Area Between Curves

Example 35:
"dx" Method

Find the area bounded by $y = x^2 + 3$, $y = -x^2 - 2$, x=-1, x=1

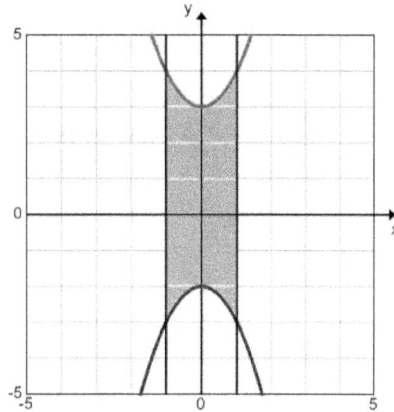

The way to find this area is to find the integral of the top function ($y = x^2 + 3$) and subtract the bottom function ($y = -x^2 - 2$). The range of x values (from -1 to 1) is the domain of this integral.

$$\int_{-1}^{1} (x^2 + 3) - (-x^2 - 2)dx$$

Simplify and combine like terms

$$\int_{-1}^{1} (x^2 + 3) + (x^2 + 2)dx$$

$$\int_{-1}^{1} 2x^2 + 5dx$$

Use the power integral rule

$$\frac{2}{3}x^3 + 5x \Big|_{-1}^{1}$$

Find $\frac{2}{3}(1)^3 + 5(1)$. Then subtract $\frac{2}{3}(-1)^3 + 5(-1)$ from that amount to find the final answer.

$$\frac{2}{3}(1)^3 + 5(1)$$

$$\frac{2}{3} + 5 = 17/3$$

93

$$\frac{2}{3}(-1)^3 + 5(-1)$$

$$-\frac{2}{3} - 5 = -17/3$$

$$\frac{17}{3} - -\frac{17}{3}$$

$$\frac{17}{3} + \frac{17}{3} = \frac{34}{3}$$

The area bounded by $y = x^2 + 3$, $y = -x^2 - 2$, x=-1, x=1 is $\frac{34}{3}$ units squared.

Example 35:
"dy" Method

Find the area between $y = \sqrt{x}$ and y=x-2 when $x \geq 0$ and $y \geq 0$

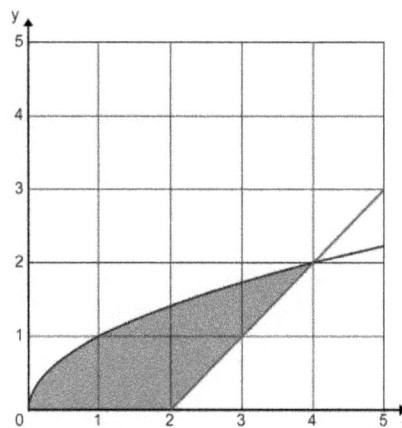

The easiest way to find this area is to express the equations in terms of y. The dx method would still find the correct area but it would be more difficult and time consuming.

$$y = \sqrt{x} \rightarrow x = y^2$$

$$y = x - 2 \rightarrow x = y + 2$$

The first step is to find the range of the integral. The overlap area is between the y values of 0 and 2 so those are the range of the integral.

Next we need to determine which equation is further to the right. In order to find the area between curves with the dy method, one must subtract the left equation from the right in the integral. In this problem $x = y + 2$ is the equation farthest to the right.

$$\int_0^2 (y+2) - y^2 \, dy$$

<u>Utilize</u> the power integral rule.

$$\frac{y^2}{2} + 2y - \frac{y^3}{3} \Big|_0^2$$

<u>Find</u> $\dfrac{(2)^2}{2} + 2(2) - \dfrac{(2)^3}{3}$

$$\frac{(2)^2}{2} + 2(2) - \frac{(2)^3}{3}$$

$$\frac{4}{2} + 2(2) - \frac{8}{3}$$

$$2 + 4 - \frac{8}{3}$$

$$\frac{6}{3} + \frac{12}{3} - \frac{8}{3} = \frac{10}{3}$$

<u>Subtract</u> $\dfrac{(0)^2}{2} + 2(0) - \dfrac{(0)^3}{3}$ from $\dfrac{10}{3}$

$$\frac{(0)^2}{2} + 2(0) - \frac{(0)^3}{3}$$

$$\frac{(0)^2}{2} + 2(0) - \frac{(0)^3}{3} = 0$$

$$\frac{10}{3} - 0 = \frac{10}{3}$$

The area between $y = \sqrt{x}$ and y=x-2 when $x \geq 0$ and $y \geq 0$ is $\dfrac{10}{3}$ units^2

36. Disks, Washers, and Shells

a. Disks

The equation to find the volume of disks revolving around the x-axis is $\pi \int_a^b (f(x))^2 \, dx$.

With disks, f(x) is the same as the radius.

Example 36a-1:

Disks—X-axis Rotation

Find the volume if the region bounded by $y = 1.5\sqrt{x}$ and the line x=3 is rotated around the x-axis.

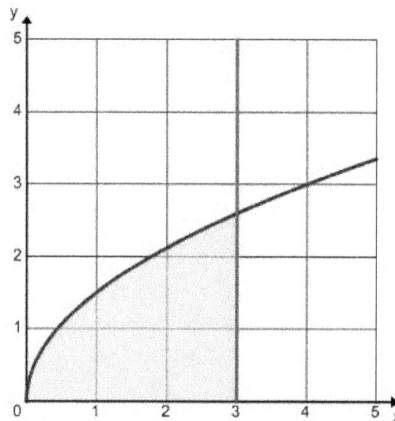

The first step is to <u>insert</u> the values we know into the disk equation $\pi \int_a^b (f(x))^2 \, dx$. b is 3 and a is zero as we can see on the graph. f(x) is $1.5\sqrt{x}$

$$\pi \int_0^3 (1.5\sqrt{x})^2 \, dx$$

Now <u>simplify</u> $(1.5\sqrt{x})^2$

$$(1.5\sqrt{x})^2 = 2.25x$$

$$\pi \int_0^3 2.25x \, dx$$

<u>Use</u> the power integral rule to find the integral. <u>Convert</u> 2.25 into a fraction.

$$\pi \int_0^3 2.25x \, dx$$

96

$$\pi \int_0^3 \frac{9}{4} x\, dx$$

$$(\pi)\frac{9}{8}x^2 \Big|_0^3$$

Find $(\pi)\dfrac{9}{8}(3)^2$

$$(\pi)\frac{9}{8}(9)$$

$$\frac{81\pi}{8}$$

Subtract $(\pi)\dfrac{9}{8}(0)^2$ from $\dfrac{81\pi}{8}$

$$(\pi)\frac{9}{8}(0)^2 = 0$$

$$\frac{81\pi}{8} - 0 = \frac{81\pi}{8}$$

The volume of the region bounded by $y = 1.5\sqrt{x}$ and the line x=3 that is rotated around the x-axis is $\dfrac{81\pi}{8}$ units cubed.

The equation to find the volume of disks revolving around the y-axis is $\pi \int_a^b (f(y))^2\, dy$. With disks, f(y) is the same as the radius.

Example 36a-2:
Disks—Y-axis Rotation

Find the volume if the region bounded by $y = x^2$ and the line y=4 is rotated around the y-axis.

97

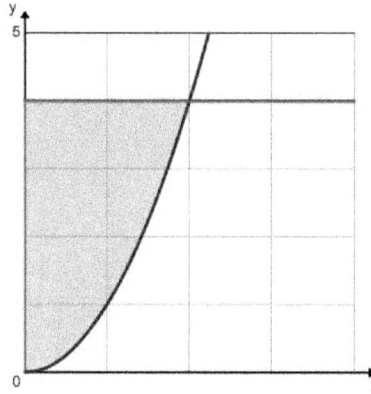

The first step here is to <u>convert</u> $y = x^2$ into terms of y since the rotation is around the y-axis.

$$y = x^2$$

$$x^2 = y$$

$$x = \sqrt{y}$$

Now <u>insert</u> the known values into the equation $\pi \int_a^b (f(y))^2 \, dy$. f(y) is $\sqrt{y}$. b is 4 and a is zero. They are not 2 and zero because this problem is in terms of y.

$$\pi \int_0^4 (\sqrt{y})^2 \, dy$$

<u>Simplify</u>

$$\pi \int_0^4 (y) dy$$

<u>Use</u> the power integral rule to find the integral.

$$\pi \int_0^4 (y) dy$$

$$(\pi) \frac{y^2}{2} \Big|_0^4$$

<u>Find</u> $(\pi) \dfrac{(4)^2}{2}$

$$(\pi)\frac{(4)^2}{2}$$

$$(\pi)\frac{16}{2}$$

$$8\pi$$

Subtract $(\pi)\dfrac{(0)^2}{2}$ from 8π

$$(\pi)\frac{(0)^2}{2}=0$$

$$8\pi - 0 = 8\pi$$

The volume of the region bounded by $y = x^2$ and the line y=4 rotated around the y-axis is 8π units^3.

b. Washers

"Washers" occur when the region rotated around either the x or y axis does not actually touch the axis. The volume of the shape appears to have a hole in the middle of it, thus leading to its name. The equation for washers rotated around the x-axis is

$$\pi\int_a^b (\text{outer radius})^2 - (\text{inner radius})^2$$

Example 36b:

Washers

Find the volume if the region bounded by $y = -x^2 + 4$, x=-1, x=1, and the line y=1 is rotated about the x-axis. This region is lightly shaded in the graph below.

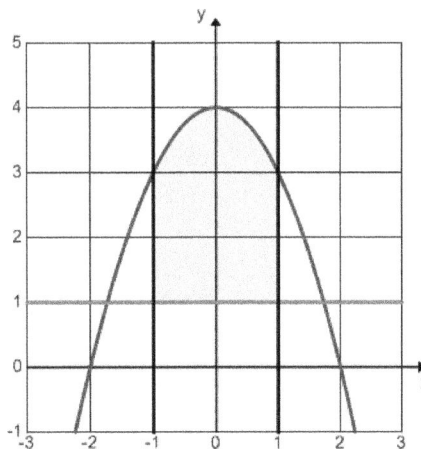

$y = -x^2 + 4$ is the outer radius in this problem and y=1 is the inner radius. This is because $y = -x^2 + 4$ is farther from the x-axis. Now <u>insert</u> the known information into the washer equation.

$$\pi \int_a^b (\text{outer radius})^2 - (\text{inner radius})^2$$

$$\pi \int_{-1}^1 (-x^2 + 4)^2 - (1)^2 \, dx$$

<u>Simplify</u>

$$\pi \int_{-1}^1 (x^4 - 8x^2 + 16) - (1) \, dx$$

$$\pi \int_{-1}^1 x^4 - 8x^2 + 15 \, dx$$

<u>Use</u> the power integral rule to find the integral.

$$(\pi)(\frac{1}{5}x^5 - \frac{8}{3}x^3 + 15x)\Big|_{-1}^1$$

<u>Find</u> $(\pi)(\frac{1}{5}(1)^5 - \frac{8}{3}(1)^3 + 15(1))$ and <u>simplify</u>

$$(\pi)(\frac{1}{5}(1) - \frac{8}{3}(1) + 15(1))$$

$$(\pi)(\frac{1}{5} - \frac{8}{3} + 15)$$

$$(\pi)(\frac{3}{15} - \frac{40}{15} + \frac{225}{15})$$

$$(\pi)(\frac{188}{15})$$

$$\frac{188\pi}{15}$$

<u>Subtract</u> $(\pi)(\frac{1}{5}(-1)^5 - \frac{8}{3}(-1)^3 + 15(-1))$ from $\frac{188\pi}{15}$

100

$$(\pi)(\frac{1}{5}(-1)^5 - \frac{8}{3}(-1)^3 + 15(-1))$$

$$(\pi)(\frac{1}{5}(-1) - \frac{8}{3}(-1) + 15(-1))$$

$$(\pi)(\frac{-1}{5} + \frac{8}{3} - 15)$$

$$(\pi)(\frac{-3}{15} + \frac{40}{15} - \frac{225}{15})$$

$$(\pi)(-\frac{188}{15})$$

$$\frac{-188\pi}{15}$$

$$\frac{188\pi}{15} - \frac{-188\pi}{15}$$

$$\frac{188\pi}{15} + \frac{188\pi}{15} = \frac{376\pi}{15}$$

The volume of the region bounded by $y = -x^2 + 4$, x=-1, x=1, and the line y=1 rotated about the x-axis $\dfrac{376\pi}{15}$ units cubed.

c. Shell Method

The shell method is often preferred for problems where the solid is in terms of x and is rotated around the y-axis (and is not touching the y-axis). This is also true if the solid is in terms of y and is rotated around the x-axis(and is separated from the x-axis). The shell method equations are $v = \int_{a}^{b} 2\pi(x)(f(x))dx$ when in terms of x and

$v = \int_{a}^{b} 2\pi(y)(f(y))dy$ when in terms of y.

Whether to use the washer and shell method is ultimately your choice when there is a hole in the rotated solid. It is best to be familiar with both options!

Example 36c:
Shell Method

Find the volume of the region bounded by $y = (x-2)(x-5)^2$, y>0, x>0, and x=5 when rotated around the y-axis.

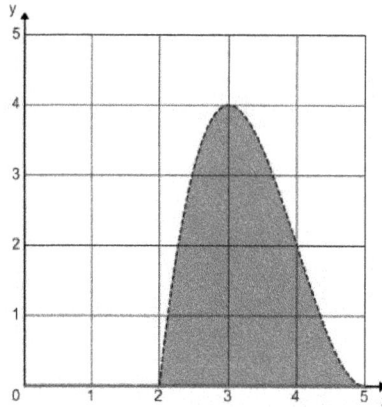

The first step is to <u>find</u> b and a of $v = \int_a^b 2\pi(x)(\mathrm{f}(x))dx$. This is done by setting $y = (x-2)(x-5)^2$ equal to zero.

$$0 = (x-2)(x-5)^2$$

When x=2 and x=5, $0 = (x-2)(x-5)^2$. Now <u>insert</u> the known information into the shell method equation (use the one in terms of x).

$$v = \int_2^5 2\pi(x)((x-2)(x-5)^2)dx$$

<u>Move</u> the 2π in front of the integral and <u>simplify</u>

$$2\pi \int_2^5 (x)((x-2)(x-5)^2)dx$$

$$2\pi \int_2^5 (x)((x-2)(x^2-10x+25))dx$$

$$2\pi \int_2^5 (x)(x^3-2x^2-10x^2+20x+25x-50)dx$$

$$2\pi \int_2^5 (x^4-2x^3-10x^3+20x^2+25x^2-50x)dx$$

<u>Combine</u> like terms

102

$$2\pi\int_{2}^{5}(x^4-12x^3+45x^2-50x)dx$$

Use the power integral rule to find the integral.

$$2\pi(\frac{x^5}{5}-3x^4+15x^3-25x^2)\Big|_{2}^{5}$$

Find $2\pi(\frac{(5)^5}{5}-3(5)^4+15(5)^3-25(5)^2)$

$$2\pi(\frac{3125}{5}-3(625)+15(125)-25(25))$$

$$2\pi(625-1875+1875-625)$$

$$2\pi(625-1875+1875-625)=0$$

Subtract $2\pi(\frac{(2)^5}{5}-3(2)^4+15(2)^3-25(2)^2)$ from 0

$$2\pi(\frac{32}{5}-3(16)+15(8)-25(4))$$

$$2\pi(\frac{32}{5}-48+120-100)$$

$$2\pi(\frac{32}{5}-\frac{240}{5}+\frac{600}{5}-\frac{500}{5})$$

$$2\pi(\frac{32}{5}-\frac{240}{5}+\frac{600}{5}-\frac{500}{5})=\frac{-108(2\pi)}{5}$$

$$\frac{-216\pi}{5}$$

$$0-\frac{-216\pi}{5}$$

$$0+\frac{216\pi}{5}=\frac{216\pi}{5}$$

The volume of the region rotated around the y axis that is bounded

by $y = (x - 2)(x - 5)^2$ and the x-axis is $\dfrac{216\pi}{5}$ units^3.

37. Volume by Cross Sections

The volume of a solid with cross sections is can be found with a definite integral. Integrate the area of each cross section to find the final answer. Knowing the formula for each cross section (semicircle, quartercircle, etc.) is crucial.

The areas will be functions of x if the cross section is perpendicular to the x-axis. Additionally, they will be functions of y if the cross sections are perpendicular to the y-axis.

Example 37:
Volume by Cross Sections--Quartercircle

Find the volume of the solid that has a base of the area between x=0, y=0 and y=-1/3x+1, if the vertical cross sections are semicircles.

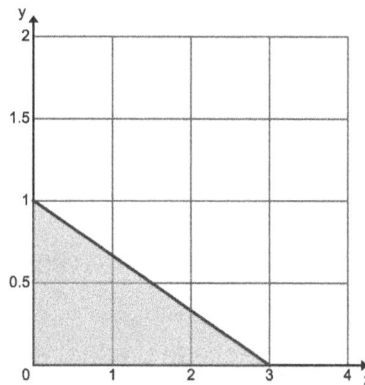

This will be a function of x since the cross sections are perpendicular to the x-axis. The equation for a semicircle's area is $A = \dfrac{\pi}{2} r^2$. The first step is to find the radius of each semicircle. The diameter of each semicircle is y=-1/3x+1. To find the radius, <u>divide</u> the diameter by 2.

$$r = \dfrac{-\dfrac{1}{3}x + 1}{2}$$

$$r = -\dfrac{1}{6}x + \dfrac{1}{2}$$

Now that the radius is known, <u>substitute</u> the radius into the semicircle area equation and <u>simplify</u>. The semicircle area equation is $\frac{\pi}{2}(r)^2$

$$A = \frac{\pi}{2}(-\frac{1}{6}x + \frac{1}{2})^2$$

$$A = \frac{\pi}{2}(-\frac{1}{6}x + \frac{3}{6})^2$$

$$A = \frac{\pi}{2}(\frac{-x+3}{6})^2$$

$$A = \frac{\pi}{72}(-x+3)^2$$

Now <u>find</u> the integral of this area. The bounds of the integral are 0 and 3 (see graph).

$$\int_0^3 \frac{\pi}{72}(-x+3)^2 dx$$

<u>Simplify</u> and <u>find</u> the volume

$$\frac{\pi}{72}\int_0^3 (-x+3)^2 dx$$

$$\frac{\pi}{72}\int_0^3 (x^2 - 6x + 9)dx$$

$$\frac{\pi}{72}\left[\frac{1}{3}x^3 - 3x^2 + 9x\right]_0^3$$

$$\frac{\pi}{72}\left[\frac{1}{3}(3)^3 - 3(3)^2 + 9(3)\right]_0^3$$

$$\frac{\pi}{72}\left[\frac{1}{3}(27) - 3(9) + 27\right]_0^3$$

$$\frac{\pi}{72}\left[9 - 27 + 27\right]_0^3 = \frac{9\pi}{72} = \frac{\pi}{8}$$

$$\frac{\pi}{72}\left[\frac{1}{3}(0)^3 - 3(0)^2 + 9(0)\right]_0^3 = 0$$

$$\frac{\pi}{8} - 0 = \frac{\pi}{8}$$

The volume of the solid that has a base of the area between x=0, y=0 and y=-1/3x+1, with vertical semicircle cross sections is $\frac{\pi}{8} \, units^3$.

38. Average Value of a Function

The average value of a function can be found between a period of time with the equation $\frac{1}{b-a}\int_a^b f(x)dx$.

Example 38:
Average Value of a Function

The flow of traffic in a certain location has been calculated to be F(t)=$t^3 - 5t + 50$ cars per minute. What is the average number of cars per minute driving through this location from $1 \le t \le 4$?

For this problem, the equation $\frac{1}{b-a}\int_a^b f(x)dx$ is needed. b is 4 and a is 1.

F(t)=$t^3 - 5t + 50$ is the function we need to <u>find</u> the integral for.

$$\frac{1}{4-1}\int_1^4 t^3 - 5t + 50 dt$$

$$\frac{1}{3}\int_1^4 t^3 - 5t + 50 dt$$

Integrate

$$\frac{1}{3}\left[\frac{1}{4}t^4 - \frac{5}{2}t^2 + 50t\right]_1^4$$

<u>Find</u> $\frac{1}{3}\left[\frac{1}{4}(4)^4 - \frac{5}{2}(4)^2 + 50(4)\right]_1^4$ and then <u>subtract</u> $\frac{1}{3}\left[\frac{1}{4}(1)^4 - \frac{5}{2}(1)^2 + 50(1)\right]_1^4$ to find the answer.

$$\frac{1}{3}\left[\frac{1}{4}(4)^4 - \frac{5}{2}(4)^2 + 50(4)\right]_1^4$$

106

$$\frac{1}{3}\left[\frac{1}{4}(256) - \frac{5}{2}(16) + 200\right]_1^4$$

$$\frac{1}{3}\left[64 - 40 + 200\right]_1^4$$

$$\frac{1}{3}\left[64 - 40 + 200\right]_1^4$$

$$\frac{1}{3}(224)$$

$$\frac{224}{3}$$

$$\frac{1}{3}\left[\frac{1}{4}(1)^4 - \frac{5}{2}(1)^2 + 50(1)\right]_1^4$$

$$\frac{1}{3}\left[\frac{1}{4} - \frac{5}{2} + 50\right]_1^4$$

$$\frac{1}{3}\left[\frac{1}{4} - \frac{10}{4} + \frac{200}{4}\right]_1^4$$

$$\frac{1}{3}(\frac{191}{4})$$

$$\frac{191}{12}$$

$$\frac{224}{3} - \frac{191}{12}$$

$$\frac{896}{12} - \frac{191}{12}$$

$$\frac{705}{12} = \frac{235}{4} = 58.75$$

58.75 cars pass the specified location per minute from $1 \le t \le 4$.

Ten Test Taking Tips
for the Calculus AB AP Test
According to Ryan

1. Complete a released version of the Calculus AB AP test for practice
2. Eat well the day of the test
3. Sleep for atleast 7-8 hours the night before if possible
4. Do not forget your calculator
5. Be confident; you only need to get about 40% of the possible points to pass
6. Always show all your work so you can get partial credit, even if final answer is wrong
7. Do not forget the +C of indefinite integration problems
8. Do not spend too much time on any one question
9. Wear a watch
10. And most importantly, use **Calculus Express** to study!

Order Form

If you liked Calculus Express, recommend it to your friends and fellow students!

Table of Contents

* Limits
* Derivatives
* Applications of Derivatives
* Integrals
* Applications of Integrals

Benefits of Calculus Express

Calculus Express is a concise, easy-to-study test preparation guide to help students improve their Calculus AB Advanced Placement (AP) exam scores. To maximize relevancy, critical content is modeled after the outline of the Calculus AB AP test promulgated by The College Board.

The primary feature of Calculus Express is that it contains all necessary information in 100+ pages. This enables you to truly cram for the test and walk into the exam site having all the key material in your short-term memory! You cannot accomplish this using competing study guides.

Contact Information of Publisher

Ryan Mettling
Performance Programs Company
6810 190th Street East
Bradenton, FL 34211
941-677-6043
ryan@performanceprogramscompany.com

www.ingramcontent.com/pod-product-compliance
Lightning Source LLC
Chambersburg PA
CBHW051415200326
41520CB00023B/7244